Advanced Structured Materials

Volume 222

Common engineering materials are reaching their limits in many applications, and new developments are required to meet the increasing demands on engineering materials. The performance of materials can be improved by combining different materials to achieve better properties than with a single constituent, or by shaping the material or constituents into a specific structure. The interaction between material and structure can occur at different length scales, such as the micro, meso, or macro scale, and offers potential applications in very different fields.

This book series addresses the fundamental relationships between materials and their structure on overall properties (e.g., mechanical, thermal, chemical, electrical, or magnetic properties, etc.). Experimental data and procedures are presented, as well as methods for modeling structures and materials using numerical and analytical approaches. In addition, the series shows how these materials engineering and design processes are implemented and how new technologies can be used to optimize materials and processes.

Advanced Structured Materials is indexed in Google Scholar and Scopus.

Vladimir Metrikin · Igumnov Leonid

Forced Oscillations of Multidimensional Highly Nonlinear Systems

Deterministic Chaos and Random Changes in Parameters

Springer

Vladimir Metrikin
Lobachevsky State University of Nizhny Novgorod
Nizhniy Novgorod, Russia

Igumnov Leonid
Lobachevsky State University of Nizhny Novgorod
Nizhniy Novgorod, Russia

ISSN 1869-8433 ISSN 1869-8441 (electronic)
Advanced Structured Materials
ISBN 978-981-97-6778-6 ISBN 978-981-97-6779-3 (eBook)
https://doi.org/10.1007/978-981-97-6779-3

This Springer imprint is published by the registered company Springer Nature Singapore Pte Ltd.
The registered company address is: 152 Beach Road, #21-01/04 Gateway East, Singapore 189721, Singapore

Preface

The monograph is devoted to the presentation of methods for studying oscillations under external periodic influence and random changes in parameters in dynamic systems with nonlinearities that have discontinuities and kinks. The analysis of dynamic systems is based on effective approaches and algorithms of the method of point mappings of Poincaré surfaces, developed by the Nizhny Novgorod scientific school of academician A. A. Andronova. Considerable attention is paid to the study of the general picture of the possible behavior of systems and its use in solving various applied problems. Using specific examples, it is shown that this approach to research allows not only to advance in the study of known nonlinear features, but also to discover new effects and applications. The presentation is illustrated by numerous specific examples of oscillatory strongly nonlinear systems with discontinuous nonlinearities or piecewise continuous nonlinearities. A separate chapter is devoted to the development and use of the point mapping method for random changes in the parameters of dynamic systems, as well as in dynamic systems subject to external seismic loads.

For scientists and engineers who turn in their activities to the theory of nonlinear oscillations and its applications, graduate students and senior students.

The book was prepared with partial support from the Ministry of Science and Higher Education within the framework of the basic part of the state assignment (project No. FSWR-2023-0036) and the development of the Scientific and Educational Mathematical Center "Mathematics for Future Technologies", as well as with partial financial support from the Russian Science Foundation (projects No. 16-19-10237, No. 22-19-00138).

Nizhniy Novgorod, Russia

Vladimir Metrikin
Igumnov Leonid

Contents

Chapter 1
Introduction

Abstract The introduction describes historical information about the state of the nonlinear theory of vibration-impact (strongly nonlinear) dynamic systems, and also lists numerous machines and mechanisms whose operation in industrial, civil, road, hydraulic, and other types of production is necessarily associated with impact interactions of elements. It is indicated that the volume of this work is so large that the problem of creating high-performance and reliable systems is extremely urgent. It is substantiated that the theoretical study of the nonlinear dynamics of multi-mass systems with impact interactions makes it natural to use the numerical and analytical method of point mappings of Poincaré surfaces. The main achievements in the theory of research of vibro-impact machines and mechanisms are presented. It is especially noted that the work was written in the traditions of the Nizhny Novgorod school of the theory of nonlinear oscillations—the school of Academician A.A. Andronov and presents the main results of the four chapters of the book.

Keywords Vibration-impact systems · Nonlinear theory · Point mapping method · Stability · Periodic motion · Random changes in parameters

Industrial, civil, road, hydrotechnical, and other types of construction are characterized by labor-intensive operations, including the so-called "zero cycle", consisting of various kinds of drilling, soil destruction or compaction, penetration of various elements into soil (piles, grooves, pipes, beams, shells, structures, etc.) and their extraction, etc. The problem of creating highly productive and reliable systems (mechanical, electrical, as well as pneumatic and hydraulic) designed to manage this workload is extremely urgent.

Among the currently existing machines and mechanisms that perform all kinds of aforementioned activities, we can distinguish vibration and vibro-impact machines. The application of vibrations alters the physical and mechanical properties of the soil and significantly facilitates the work cycle. In addition, vibro-impact machines generate additional impulses and exhibit a relatively high immersion velocity. Vibration and vibro-impact machines and devices are also widely employed in industry, mining, national economy, medicine and domestic service.

V. Metrikin and I. Leonid, *Forced Oscillations of Multidimensional Highly Nonlinear Systems*, Advanced Structured Materials 222,
https://doi.org/10.1007/978-981-97-6779-3_1

The extensive application of vibration and vibro-impact mechanisms enables studying their main characteristics and determining parameters that ensure optimal working cycle. For this, it is necessary to study all possible motion modes for different set of parameters that are implemented in the considered systems.

Currently, there are fairly complete studies of vibro-impact systems with one degree of freedom, while the complete study of dynamic multidimensional systems with impact interactions is not yet possible. Therefore, the study of such systems is often reduced to studying their behavior in the most important motion modes, in particular, periodic ones.

From the standpoint of the theory of oscillations, systems with impact interactions are highly nonlinear systems. When studying the dynamics of various vibro-impact systems and mechanisms in the first approximation, we may assume that the collision time is negligible compared to the oscillation period. Therefore, when taking into account impact interaction, the impact process itself may not be considered, but one can employ the pulse theorem and the coefficient of restitution upon impact. Taking into account, impact interactions described by Newton's well-known hypothesis lead to discontinuous and instantaneous change in velocity in these systems (in some configurations) at the moments of impacts. In phase space, this corresponds to the presence of hypersurfaces which divide the entire phase space into parts. When the phase point gets onto one of these surfaces, it exhibits instant and stick–slip motion along it. A large group of technical devices also have such phase space structure: clockwork mechanisms; systems associated with vibroimmersing, vibrotransport and vibroseparation; pneumatic impact mechanisms; vibromills and test benches; shock absorbers; mechanisms associated with cyclic automatics, onset impact interactions due to the presence of gaps; as well as circuit-opening electrical systems; dynamical systems with a small parameter at the highest derivative and many others.

The impacts applied to the system as a whole or to its elements and high work rate make the well-known approximate methods (harmonic linearization method, small parameter method, averaging method) hardly acceptable for studying vibro-impact systems and, on the contrary, the method of point mappings becomes more natural in terms of application.

A large number of papers have been devoted to the study of various dynamic systems with impact interactions. The first theoretical investigations of such systems are related to the simplest single-mass mechanism in works by A. A. Andronov, A. A. Witt, S. E. Khaykin, N. N. Bautin, Yu. I. Neymark, Z. M. Axelrod and others (see [1, 2] and bibliography).

The next followed after studied in detail clockwork mechanism was the system with impact interactions, namely vibro-impact mechanism which is a mass on a spring subjected to an external periodic force, the oscillations of which are limited by a fixed plane.

The forced oscillations of mass impacting against a fixed limiter were studied for the first time by I. G. Rusakov and A. A. Kharkevich based of the existence of stable periodic motions determined by means of usual stitching method [3–18].

Later, the fitting method was used in studying various dynamic systems with impact interactions by Kobrinsky A. E., Brunshteyn R. Ye., Babitsky V. I. and many others [19–34].

Lifshits and Rudolf [35, 36] studied the dependence of impact pulse on parameters of a single-mass vibro-hammer, and the results of which were compared with experimental data.

Due to a wide application of a vibro-hammer for driving a groove, the forced oscillations of a single-mass vibro-impact mechanism were subjected to theoretical research. Tsaplin in [37] showed that vibro-impact mechanism can have periodic motions both with a periodic external force and with a period multiple to the period of external effect.

Further, Barkan and Schechter [38] take into account the possibility of additional impact during the motion period of a vibro-impact mechanism. As a result, the authors noted that a vibro-hammer with an absolutely inelastic impact can have periodic motions with prolonged stops. At the same time, the authors considered two-mass models. This was due to the fact that having a simple design, easy and convenient tuning, single-mass vibro-impact mechanisms had some drawbacks. First, it was impossible to control the vibro-hammer weight. Second, there were no normal operation conditions for a vibrato which serves as an impact part of the mechanism.

New models of two-mass vibro-impact mechanisms [39–41] have been created by constructors and engineers. In the first designs of Savinova and Eliseev [41], it was possible to easily change the weight of the system and the vibrator still remained the impact part of the mechanism. In the models proposed by D. D. Barkan and A. Ya. Luskin, a special impact mass separates the vibrator from the head-cap by means of springs. The use of a special impact mass in the vibro-impact mechanism considerably extended the service life of vibrators.

Two-mass vibro-impact machines have a more complex structure and require more difficult tuning to the optimum mode of operation. All above mentioned set before the researchers new tasks on studying the forced oscillations of two-mass systems with impact interactions.

In the work [42], D. D. Barkan and O. Ya. Schechter considered forced oscillations of the vibrator with a movable limiter using the fitting method, and in [43], they give consideration to a two-mass vibro-impact mechanism with a special impact mass.

In all the aforementioned works, the main attention is paid to deriving the conditions for existence of periodic modes of motion that do not really exist due to the loss of stability.

The first attempts to investigate the stability of periodic motions of the vibro-impact mechanism were made by Osmakov in [44]. The author proposes to study the stability at each point of the parameter domain by incrementing the initial conditions and comparing them with increments through n- periods of motions.

Further, Sparenber in [45] obtained stability conditions for a single-stage vibro-impact mechanism with the coefficient of restitution at impact equal to one.

A detailed study of a single-mass vibro-impact mechanism was carried out by Bespalova [46]. In this work, for studying the dynamics of the vibro-impact mechanism, the method of point mappings [47–51] was applied, which was developed for the problems of nonlinear vibrations by Andronov's Gorky school. The method of point mappings, as known, is based on the possibility of reducing the problem of investigation of dynamical systems to studying the properties of some point mappings of Poincare surfaces and, in particular, of the invariant points of these mappings, which correspond to periodic motions and equilibrium conditions of the systems under study.

The application of this method for studying the dynamics of a single-mass vibro-impact mechanism enabled the author to find all possible types of periodic motions and investigate their stability. As a result of the research, the author classified all kinds of periodic motions. Hence, it follows that any periodic motion of a single-mass vibro-impact mechanism can be characterized by two numbers m and n, where m is the number of impacts per motion period, n is the multiplicity of the period of motion to that of external force. Investigations of the stability of single-impact n-fold periodic motions showed that some found earlier [2] domains of existence disappear due to the loss of stability. In addition, the other part of the domains of existence disappears due to the presence of a bifurcation boundary separating periodic motions with a different number of impacts per period of motion.

For absolutely inelastic impact, the application of point mappings method reduced in this case to transformation of a circle into itself, enabled the author to obtain a numerical partition of the plane of system parameters into domains of various periodic motions. From this, it is seen that even in such a simple model, a large variety of different types of periodic motions both with and without stops are possible.

Further, the method of point mapping and the theory of bifurcations [49–51] were widely applied for investigation of dynamics of various systems with impact interactions as presented in the works of Feigin [52–58], Osmakova and Savinova [59–61], Ragulskisa et al. [62–64], Bykhovsky [65–67], Zaretsky [65–69], Peterka [70, 71], Lukomsky [72], Goldin [73, 74] and many others (see, for example, [62, 67, 75] and the bibliography).

It should be noted that the use of the method of point mapping requires quite a lot of computational work. At present, this obstacle hindering to carry out specific research (in early studies of vibrational immersion [76–87]) can be overcome with the help of computer experiments.

Recently, the widespread use of the results of computer experiments for investigation the dynamics of various systems with impact interactions made it possible to solve a number of important practical problems, and the solution of which was delayed due to cumbersome computational work. The task of research into stability of multi-mass vibro-impact systems was set by Yu. I. Neymark back in the 50 s but due to computational difficulties remained unresolved for a long period of time.

In this regard, it should be noted that Batalova [88–90] presents the algorithms for constructing and studying point mapping of a straight line into a straight line and a plane into a plane as applied to electronic digital computers.

Unable to provide a complete overview of the available research results on specific dynamic systems with impact interactions by the method of point mapping, we have to confine ourselves to indicating a number of practically important problems.

The works of Feigin [52–54, 75, 91] are devoted to studying the dynamics of impact absorbers and systems of two colliding masses forming a gap. The works of Shlyakhtin [92–94] aimed at studying the periodic motions of vibro-impact systems with an internal combustion engine, while Boyarinova and Neymark [95] clarify the conditions for the onset of radial vibrations with impacts in a ball bearing and Podchufarov [96–100] investigated the dynamics of some models of cyclic automation, etc. [101–105].

Dynamic systems that are nonlinear in the intervals between impacts are more complex impact interaction systems. Therefore, the application of accurate methods for studying the dynamics of such systems is extremely difficult and, in many cases, impossible.

In view of this, in [62, 69, 106–113] and a number of other works, it was necessary to use either accurate methods in combination with approximate or only approximate ones. In fundamental works of Neymark et al. [114–117], the method of point mapping in combination with the small parameter method was developed in order to study oscillations in nonlinear systems.

Nachaev [106] studied the dynamics of a self-synchronizing vibro-impact grinder by the method of point mappings with application of small parameter methods. Zaretsky [69] considered the forced oscillations of a single-mass vibration striker, and the motion of which in the intervals between impacts was described by a system of nonlinear equations. Approximate methods of Krylov-Bogolyubov-Mitropolsky, Galerkin, etc., were applied in the works [62, 107–109, 112, 113] and others [118–120].

To study the dynamics of systems with impact interactions, electronic modeling is widely used on continuous machines (computers). Although continuous machines have less accuracy as compared to digital ones (digital computers), they allow in a relatively short time to qualitatively determine and evaluate the possible types of both periodic and non-periodic motions.

A large number of works are devoted to the study of systems with impact interactions by means of computers (see references [46, 62–68, 70, 71, 73, 74, 121–126] and others).

In brief review, we considered papers devoted to studying the dynamics of nonlinear oscillations for deterministic parameter values. It is obvious, that in all real systems, including systems with impact interactions, a number of parameters vary randomly.

The oscillations in systems with impact interactions with randomly varying parameters have been much less studied as compared with deterministic investigations. In 1933, A. A. Andronov, A. A. Witt, and L. S. Pontryagin in the fundamental work [127] pointed out the necessity for statistical approach to dynamic systems.

Currently, there exist quite a lot of works on statistical research in radio engineering, electronics, and other fields of science and technology (see monographs [128–130] and the literature cited therein).

Rytov's work [131] was the first work on research into fluctuations caused by small and short-correlated random forces in piecewise linear self-oscillating systems using the method of point mappings. The work presents statistical characteristics of fluctuations (in particular, the "phase" diffusion coefficient). As an example, a tube generator with a Z-characteristic is considered. Further, the proposed method for finding statistical characteristics was generalized by Sadekov [132] in case of non-delta-correlated noises affecting a piecewise linear system with an arbitrary number of degrees of freedom.

There are also a number of works on statistical study of various dynamical systems, e.g., the works of Blekhman [133, 134], Dimmetnberg [135–137], Kolovsky [113], and many others [138–146].

In early 60 s, the works on research into dynamics of vibro-impact systems started to appear. Kolovsky and Babitsky [147, 148] examined the oscillations of a linear system with a limiter at random loads for the speed recovery coefficient upon impact equal to unity, using exact and approximate methods.

The works of Zaretsky et al. [149, 150] are devoted to some statistical studies of motions of vibro-impact mechanism mainly using analog techniques.

Kobrinsky [151, 152] found the statistical characteristics of a single-mass system with impact interactions for small random deviations of the coefficient of restitution during impact using the fitting method.

The works ([153–163] and others) deal with studying the stochastic stability of systems under random parameter perturbations. Weidenhammer [153] formulated stability conditions for equilibria of a system with two degrees of freedom parametrically excited by a stationary Gaussian white noise. For stochastic stability, the first and second statistical moments of coordinates and velocities are required to remain limited over time.

In the works of Khasminsky [154, 155, 157, 159] and M. B. Nevelson [155–158], various definitions of stochastic stability which are a generalization of stability concepts in the sense of Lyapunov are given. However, interesting results obtained using these definitions mainly relate to linear systems or to systems that are close to linear. In the works of Brusina and Tai [161], an attempt is made to introduce the concept of stochastic stability of stochastic systems which are strongly nonlinear systems by generalizing the known V.M. Popov's method.

This monograph is devoted to the theory of various vibro-impact systems (two-mass and multi-mass), to the study of the effect of viscous friction forces on forced oscillations of the systems with impact interactions (single-mass and two-mass), as well as to the study of the motions of various single-mass vibro-impact mechanisms with a random change of parameters.

The solution to these problems can be divided into four parts:

1. Research into the theory of periodic motions and their stability, construction of C-bifurcations of periodic motions and their dependence on parameters in two-mass vibro-impact systems.
2. Study of periodic motions and their stability in a multi-mass vibro-impact mechanism.
3. Study of the effect of viscous friction on the stable operation of single-mass and two-mass vibro-impact systems.

4. Study of the dynamics of vibro-impact systems with random changes in parameters.

Each of these tasks is of independent interest.

The study of these four problems is carried out by the method of point mappings [47–51] using electronic digital and analog machines.

The second chapter deals with the dynamics of various two-mass models of a vibro-striker with a fixed limiter. The study starts with a common two-mass model consisting of two masses m_1 (impact mass) and m_2 (vibrator), fixed by means of springs C_1 and C_3, connected by a connecting spring C_2. An external harmonic force acts on the upper mass m_2. . The amplitude of oscillations of the lower mass is limited by a fixed barrier (limiter) in contact with which there is an instant elastic impact, characterized by the coefficient of restitution R.

Under simplifying assumptions made in § 2.1, the equations of motion of the system in the intervals between impacts are written in the form:

$$\begin{aligned} m_1\ddot{x}_1 + (C_1 + C_2)x_1 - C_2x_2 &= P_1 \\ m_2\ddot{x}_2 + (C_2 + C_3)x_2 - C_2x_1 &= P_2 + F\ \sin\omega t, \end{aligned} \tag{1.1}$$

and at the moment of contact of the lower mass with a fixed limiter

$$\dot{x}_1^+ = -R\dot{x}_0^- \tag{1.2}$$

where x_1, x_2 are displacement so flower and upper masses, $\dot{x}_1^+, \dot{x}_0^-$ are lower mass velocity immediately after and before the impact, P_1 is impact weight, P_2 is a constant force, acting on mass m_2, F, ω are the amplitude and frequency of the driving force. The equations of motion of various two-mass models considered in §§ 2.3–2.5 are obtained from the original system (1.1)–(1.2) for the following values of the spring stiffness coefficients: $C_3 = 0$, or $C_1 = 0$, or $C_1 = C_3 = 0$.

In § 2.1, the problem the forced vibrations of a vibro-impact system is reduced to studying the point mapping of a four-dimensional hyperplane (corresponding to the limiter plane) into itself in a five-dimensional phase space. The structure of the phase space is considered and the possible types of impact motions in the case of both elastic ($R \neq 0$) and inelastic ($R = 0$) impacts are indicated. It is shown that (as in the single-mass vibro-impact system [46]) any periodic motion (for $R \neq 0$) can be characterized by two numbers m and n, where m is the number of impacts per period of motion, n is the multiplicity of the period of motion to the period of the external force.

In § 2.2, in the parameter space, the coordinates of the fixed points corresponding to the periodic single-impact n-fold mode of motion are found in the parametric form. The boundaries of the existence of these periodic motions are determined.

Based both on point mappings and bifurcation theories, the equations of bifurcation surfaces are given that correspond to the violation of existence and stability conditions of fixed points (N-surfaces) and to the exit of a closed phase trajectory beyond the domain of the studied point transformation (C-surface).

Bifurcation C-surfaces, separating single-impact periodic motions from multi-impact, correspond to the rise of additional impacts during the motion period. It is noted that for parameter values at which the gap is greater than the amplitude of the forced vibrations, along with unstressed and quasiperiodic motions the motions with hard onset are possible [1].

In § 2.3, we study the dynamics of a spring-loaded two-mass vibration striker, obtained from a general two-mass model assuming the stiffness of the upper springs to be zero. This model was first considered by Barkan and Shekhter [43]. However, the authors proceeded from the existence of stable periodic motions and determined them using the usual stitching method and bifurcation *N*-, *C*-surfaces were not constructed. In this section, it is shown that the regions of existence of single-impact *n*-fold periodic motion regimes found in [43] are significantly narrowed due to their loss of stability.

For partition of the parameter space into stability regions, a fourth-degree characteristic polynomial is obtained, from which bifurcation N-surfaces for various parameter values are constructed. The most significant among domains of stability are those located near the resonant frequencies of the corresponding linear system without a limiter.

The dependence of the location of the stability regions on the parameters was studied. With an increase in the magnitude ϕ (the ratio of the weight of the upper mass to the lower), the stability regions are found to get narrow in the pre-resonance gap and to expand in the post resonance region. In this case, the stability regions shift toward the side of decreasing external force frequency ω. It is shown that with increasing ζ (the ratio of the stiffnesses of the lower springs to the upper ones), the expansion of the stability regions and a shift of the latter toward smaller ones are observed. With an increase in the coefficient of restitution R, the stability regions become more indented, and the incisions appear from negative gaps.

The study of the location of bifurcation *C*-surfaces showed that they cut off a significant part of the *N*-stability regions. A technique for constructing both *N*-stability regions and *C*-bifurcation surfaces is given. The results of electronic modeling on an MN-8 computer are presented in this section. The results agree well with the theoretical ones.

In § 2.4, we studied the model of a two-mass non-spring-loaded vibro-striker obtained from the general model, assuming $C_1 = 0$.

The stability of single-impact n-fold periodic motions is investigated. The boundaries of the stability regions of these motions are found and bifurcation C-surfaces are constructed. The regions with stable single-impact n-fold periodic motions are slightly larger than those considered in § 2.3. The dependence of the location of these regions on system parameters is investigated.

In § 2.5, stable single-impact n-fold periodic motions for a two-mass vibro-impact non-spring-loaded compactor, obtained from the general model at values $C_1 = C_3 = 0$ are found and studied. It is shown that the stability regions of periodic motions are narrow strips that expand with increasing the coefficient of restitution.

For absolutely elastic impact ($R = 1$), some general properties of bifurcation N-surfaces are given, and a technique for constructing bifurcation surfaces is indicated. The location of stability regions in three-dimensional parameter space is studied.

The chapter ends with general conclusions.

The third chapter deals with the study of a multi-mass vibro-impact system, a generalization of a two-mass system, consisting of N identical masses connected by springs of the same stiffness. Under these assumptions, it was possible to compactly find the coordinates of the fixed points corresponding to one-impact n-fold periodic motions and study their stability for a different number of masses. The problem of studying the stability of fixed points is reduced to determining the roots z_i of a characteristic polynomial $\chi(z) = 0$, which satisfies the condition $|z_i| < 1$ equivalent to finding the maximum eigenvalues modulo of the matrix A_{2N} inside a circle of unit radius. The optimal values of the system parameters are found for which the maximum eigenvalue modulo of the matrix A_{2N} (or maximum modulo root of the characteristic polynomial $\chi(z) = 0$) takes minimum value.

Chapter 3 deals with studying the effect of viscous friction forces on the stable operation of vibro-impact systems.

In § 3.1, the dynamics of a single-mass vibration strikers considered with account for viscous friction proportional to the first degree of velocity. In 3D phase space of the system, investigation of vibro-striker motions is reduced to studying point mapping of cylinder surface into itself as in the case when viscous friction is not accounted for [46].

The coordinates of the fixed points corresponding to single-impact n-fold periodic motions are obtained, and their stability is investigated. The parameters of bifurcation N-surfaces, as well as bifurcation of C-surfaces, were found and constructed in space.

It is shown that viscous friction forces when accounted for lead to smoothing infinite discontinuities of individual parts of stability regions. In this case, the magnitude of the impact velocity decreases. Some new properties of bifurcation C-surfaces are indicated, suggesting that the existing superposition of single-impact periodic motions with different multiplicities ($n = 1, 2, 3, \ldots$) is eliminated by these bifurcation surfaces. Recommendations for choosing optimal parameter values are given.

In the case of resonance, the three-dimensional space of parameters was completely partitioned into stability regions of single-impact n-fold periodic motions. It is shown that in this case, there are no multiple ($n \geq 2$) stable single-impact periodic motions. Surfaces of equal impact velocities are constructed.

In the second section, we study the operation stability of two-mass vibro-impact mechanism considered in § 4.3 with account for viscous friction forces. The developed software product enabled us to determine the coordinates of fixed points, to compute a characteristic polynomial and investigate the stability. The effect of all parameters on the location of stability regions was studied, and the impact velocity values were compared. The charter ends with main conclusions.

Chapter 4 deals with the dynamics of single-mass vibro-impact systems at small random changes of the coefficient of restitution R and the gap ξ_0 from those fixed values for which a stable periodic regime exists. In § 4.1, the statistical characteristics are determined for the cases of delta- and non-delta-correlated changes in the coefficient restitution and the gap. The parameters are determined at which small random deviations of the coefficient of restitution do not affect single-impact mode of motion. The parameters were also determined at which non-delta-correlated changes have the same effect on the vibro-impact mechanisms delta-correlated changes of R and ξ_0.

In § 4.2, the concept of stochastic stability is introduced on the basis of already existing definitions [128, 153, 154, 161]. Based on the introduced definition, one can judge the degree of stochastic stability of the considered motion, hereof, some practical recommendations for choosing the system parameters were made.

In § 4.3, we study the effect of random changes in the coefficient of restitution R by the value of the optimal impact velocity $\dot{\xi}_0^*$.

It was found that the optimal value $\dot{\xi}_0^*$ decreases with small random deviations of R, and the optimum shifts for positive gaps to smaller ω and for negative gaps to higher frequencies of the driving force. It is shown that for small random deviations of the coefficient of restitution R, the plane $1/\lambda$, ξ_0 is divided into three domains D_1, D_2, D_3, and in each of the domains, the magnitude of the average impact velocity either increases (domains D_1, D_3) or decreases (domain D_2) as compared with the velocity $\dot{\xi}_0$ in deterministic case.

§ 5.4 deals with the question of the probability of vibro-striker $P_в$ to come out of a given single-impact mode of motion. Top estimates of $P_в$ for outcome probability are found in dependence on the system parameters.

In § 5.5, the problem on vibro-striker motion at random change in R was digitally simulated. The simulation results are in good agreement with the theoretical results for deviations of the coefficient ranged from −0.05 to 0.05. The approximate value for infinitesimal of random changes is found.

Thus, the main results of the work are as follows:

1. A technique is proposed for partitioning the parameter space of vibro-impact systems with two degrees of freedom into stability regions of single-impact n-fold periodic modes of motion as applied to a computer.
2. In the parameter space, stability regions of single-impact periodic motions of different multiplicity of two-degree vibro-impact mechanisms are found and practical recommendations for choosing the optimal parameter values are given.
3. The stability of the periodic motions of a multi-mass vibro-impact system is investigated and the change in the optimal parameters depending on the number of masses is determined.
4. The effect of viscous friction forces on the stability of single-mass and two-mass vibro-impact systems is studied. It is shown that viscous friction forces decrease both the dimensions of stability regions of single-impact n-fold periodic motions and the optimal impact velocity.
5. Some new properties of bifurcation C-surfaces have been discovered. It is shown that in separate plane sections of the parameter space (and for the resonance casein the entire 3D space), bifurcation C-surfaces eliminate the imposition of periodic motion modes of various multiplicities.
6. An electronic simulation was carried out on a continuous computer MN-8 of a two-mass vibro-hammer. Simulation results are compared with theoretical ones.
7. The dynamics of single-mass vibro-impact mechanisms was studied for small random deviations of the coefficient of restitution R and the gap ξ_0:

(a) statistical characteristics are found for delta- and non-delta-correlated changes of R and ξ_0, as well as parameter values at which small random changes of R and ξ_0 do not lead to a violation of the single-impact mode;
(b) it is shown that small random deviations of R lead to a decrease in the average optimal impact velocity;
(c) an upper estimation of probability of the system outcome from a given single-impact motion mode is determined;
(d) a digital simulation was carried out on the M-20 digital computer of the vibro-striker at random change in R, as a result, the "smallness" of deviations of the coefficient of restitution R was determined.

The results were published in [164–172] and reported at the 2nd and 3rd All-Union symposia on the dynamics of machines with impact interactions at the Institute of Mechanical Engineering of the Academy of Sciences of the USSR, at the inter-university scientific conference on electric impact machines in Novosibirsk, at the All-Union Symposium on Applied Mathematics and Cybernetics in Gorky.

References

1. Andronov, A.A., Witt, A.A., Khaikin, S.E.: Theory of Oscillations. Fizmatgiz, M. (1959)
2. Ibrahim, R.A.: Vibro-impact dynamics. Modeling, Mapping and Applications vol. 43, p. 297. Springer Berlin Heidelberg (2009). https://doi.org/10.1007/978-3-642-00275-5
3. Rusakov, I.G., Kharkevich, A.A.: Forced vibrations of a system hitting a limiter. ZhTF **22**(11–12) (1942)
4. Naldi, R., Sanfelice, R.G.: Passivity-based control for hybrid systems with applications to mechanical systems exhibiting impacts. Automatica (2013). https://doi.org/10.1016/j,automatic,2013.01.018
5. Bartolo, D., Boudaoud, A., Narcy, G., Bonn, D.: Dynamics of non-Newtonian droplets. Phys. Rev. Lett. **99**, 174502 (2007)
6. Brogliato, B.: Some perspectives on the analysis and control of complementarity systems. IEEE Trans. Autom. Control **48**(6), 918–935 (2003)
7. Brogliato, B., Lozano, R., Egeland, O.: Dissipative systems analysis and control. Springer (2007)
8. Brogliato, B., Rio, A.Z.: On the control of complementary-slackness juggling mechanical systems. IEEE Trans. Autom. Control **45**(2), 235–246 (2000)
9. Carloni, R., Sanfelice, R.G., Teel, A.R., Melchiorri, C.: A hybrid control strategy for robust contact detection and force regulation. In: Proceedings of the American Control Conference, pp. 1461–1466. New York City, USA (2007)
10. Galeani, S., Menini, L., Potini, A., Tornambй, A.: Trajectory tracking for a particle in elliptical billiards. Int. J. Control. **81**(2), 189–213 (2008)
11. Goebel, R., Sanfelice, R.G., Teel, A.R.: Hybrid dynamical systems. IEEE Control. Syst. Mag. **29**(2), 28–93 (2009)
12. Goebel, R., Sanfelice, R.G., Teel, A.R.: Hybrid Dynamical Systems Modeling, Stability, and Robustness. Princeton University Press (2012)
13. Goebel, R., Teel, A.R.: Solutions to hybrid inclusions via set and graphical convergence with stability theory applications. Automatica **42**(4), 573–587 (2006)
14. Goebel, R., Teel, A.R.: Zeno behavior in homogeneous hybrid systems. In: Proceedings of the 46th IEEE Conference on Decision and Control, pp. 2758–2763. Cancun, Mexico (2008)

15. Haddad, W.M., Hui, Q.: Energy dissipating hybrid control for impulsive dynamical systems. Nonlinear Anal. **69**, 3232–3248 (2008)
16. Leine, R.I., Heimsch, T.F.: Global uniform symptotic attractive stability of the non-autonomous bouncing ball system. Physica D **241**(22), 2029–2041 (2012)
17. Leine, R.I., van de Wouw, N.: Stability properties of equilibrium sets of nonlinear mechanical systems with dry friction and impact. Nonlinear Dyn. **51**(4), 551–583 (2007)
18. Morărescu, I.C., Brogliato, B.: Passivity-based switching control of flexible-joint complementarity mechanical systems. Automatica **46**(1), 160–166 (2010)
19. Kobrinsky, A.E.: Mechanisms with Elastic Links, p. 390. M., Nauka (1964)
20. Babitsky, V.I., Krupenin, V.L.: Oscillations in Strongly Nonlinear Systems, p. 320. M., Nauka (1985)
21. Zaretsky, L.B.: Problems of the dynamics of shock-vibration machines. Sat. Mach. Mech. (17–18) (1969)
22. Kobrinsky, A.A., Kobrinsky, A.E.: Vibroimpact Systems, p. 592. M., Nauka (1973)
23. Kobrinsky, A.A., Kobrinsky, A.E.: Vibroimpact Systems (Dynamics and Stability). M., Nauka (1973)
24. Kobrinsky, A.A., Kobrinsky, A.E.: Two-Dimensional Vibroimpact Systems, p. 336. Nauka (1981)
25. Krupenin, V.L.: Mathematical modeling of dynamic systems interacting through collisions. Fundam. Res. (12–1), 167–168 (2007)
26. Krupenin, V.L.: Problems of mathematical analysis of systems. Mod. Sci.-Intensive Technol. (9), 94–95 (2009)
27. Brunshtein, R.E., Kobrinsky, A.E.: On the stability of periodic motions of vibroimpact systems. Izv. USSR Acad. Sci., OTN, Mech. Eng. (5) (1860)
28. Brunstein, R.E., Kobrinsky, A.E.: Dynamics and stability of two-mass vibroimpact systems. Izv. USSR Acad. Sci., OTN, Mech. Eng. (5) (1964)
29. Kobrinsiy, A.E., Tyves, P.I.: Dynamics and stability of systems containing two shock pairs. Mach. Sci. (4) (1965)
30. Babitsky, V.I., Kobrinsky, A.E.: Electrodynamic damper. Izv. Acad. Sci. USSR, Mech. Eng. (3) (1962)
31. Babitsky, V.I., Kobrinsky, A.E.: Periodic motions of a two-mass oscillatory system in a cavity. Proc. Semin. TMM, Nauka, M (104) (1964)
32. Babitsky, V.I., Izrailovich, M.Y.: On optimal motions of vibratory systems. Mach. Sci. (6) (1967)
33. Babitsky, V.I., Izrailovich, M.Y.: On the optimal motions of vibroimpact systems. Mach. Sci. (3) (1968)
34. Babitsky, V.I., Izrailovich M.Y.: Some questions of optimization of vibroimpact systems. Mach. Sci. (3) (1969)
35. Lifshits, P.S.: On the question of forced oscillations of a system hitting a limiter. JTF **22**(11–12) (1942)
36. Idebold, R.: Die Bewegungsvergange an elektrische angetriebenen Abbauhammern (Theorie der Kwibel-Feder-Hammer) Freiberger Forschungesch. A.2. (1953)
37. Tsaplin, S.A.: Vibroimpact mechanisms. Avtotransizdat, M (1953)
38. Barkan, D.D., Shekhter, O.Y.: On the theory of forced oscillations of a vibrator with a limiter. ZhTF **25**(13) (1955)
39. Aidanpää, J.O., Gupta, R.B.: Periodic and chaotic behavior of a threshold–limited two–degree–of–freedom system. J. Sound Vib. **165**(2), 305–327 (1993)
40. Axisa, F., Izquierdo, P.: Experiments on vibro–impact dynamics of loosely supported tubes under harmonic excitation. In: ASME Symp. on Flow–Induced Vibration and Noise, Anaheim, CA, PVP, vol. 242(2), pp. 281–299 (1992)
41. Savinov O.A., Luskin A.Y.: Vibration method of piling and its application in construction. Gosizdat, L-M (1960)
42. Barkan, D.D., Shekhter, O.Y.: Forced vibrations of a vibrator with a movable limiter. JTF **25**(13) (1955)

43. Barkan, D.D., Shekhter, O.Y.: Vibroimpact mechanisms with a special impact mass. Proceedings of the Research Institute of Foundations and Foundations, collection No. 28. Gosstroy Publishing House (1956)
44. Osmakov, S.A., Savinov, O.A.: Forced oscillations of a system of two masses interacting with the force of dry friction. Izv. Univ. Eng (4), 126–134 (1963)
45. Masri, S.F., Caughey, T.K., Sparenber, J.A.: On the stability of the impact damper. Trans. ASME. J. Appl. Metch. **33**, 586–592 (1966)
46. Bespalova, L.V.: To the theory of vibro-impact mechanism. Izvestia Acad. Sci. USSR, OTN (5) (1957)
47. Andronov, A.A., Bautin, N.N.: Movement of a Neutral Aircraft Equipped with an Autopilot and the Theory of Point Transformations of Surfaces. DAN, USSR **43**(5) (1944)
48. Andronov A.A. "Collection of Works", Ed. Academy of Sciences of the USSR, M., 1956.
49. Neimark, Y.I.: The Method of Point Transformations in the Theory of Nonlinear Oscillations. Izv. High. uch. Head, Radiophys. (1) (1958)
50. Neimark, Y.I.: The method of point mappings in the theory of nonlinear oscillations. Izv.vysh.uch. Head., Radiophys. **2**(2) (1958)
51. Neimark, Y.I.: Method of point transformations in the theory of nonlinear oscillations. Izv. High. uch. Head, Radiophys. **3**(5–6) (1958)
52. Feigin, M.I.: On the theory of nonlinear dampers. Izv. High. uch. Head, Radiophys. **2**(4) (1959)
53. Feigin, M.I.: On forced vibrations of two masses articulated with gaps. Izv. USSR Acad. Sci., OTN, Mech. Eng. (5) (1960)
54. Feigin M.I. "Resonant properties of a dynamical system with impact interactions". PMM, v.30, issue 5, 1966
55. Feigin, M.I.: On the construction of the stability region in the space of parameters of a dynamical system. PMM **32**(1) (1968)
56. Cheng, C.C., Wang, J.Y.: Free vibration analysis of a resilient impact damper. Int. J. Mech. Sci. **45**(4), 589–604 (2003)
57. Cheng, J., Xu, H.: Inner mass impact damper for attenuating structure vibration. Int. J. Solids Struct. **43**(17), 5355–5369 (2006)
58. Cheng, J., Xu, H.: Periodic motions, bifurcation, and hysteresis of the vibro–impact system. Mech. Based Design of Struct. Mach. **35**(2), 179–203 (2007)
59. Osmakov, S.A.: Forced shock oscillations and their stability. S,. Scientific Work, Tomsk Engineering and Construction Institute, vol. 3 (1957)
60. Osmakov, S.A., Savinov, O.A.: On the motion of a vibrator freely supported by an elastic limiter. Izv. USSR Acad. Sci., OTN, Mech. Eng. (3) (1960)
61. Osmanov, S.A., Savinov, O.A.: Elements of the theory and selection of parameters of free vibrohammers. Tr. in-ta / VNII hydro mechanization, dignity. Tech. Spec. Constr. Work (17) (1964)
62. Ragulkis, K.M., Vitkus, I.I., Ragulskene, V.L.: Self-synchronization of mechanical systems. Self-Synchronization Vibro-Impact Syst. Mintis, Vilnius (1965)
63. Ragulskene, V.L.: Issues of the dynamics of vibroimpact systems in periodic and transient modes. Abstr. Dissertation Candidate, Riga (1965)
64. Ragulskene, V.L., Satkevechus, E.E.: Electronic modeling of multimass vibroimpact systems. Tr. High. uch. Head Lit.SSR, Vibrotechnics (1) (1967)
65. Bykhovsky, I.I., Dorohova, A.D., Zaretsky, L.B., Lukomsky, S.I.: On some periodic motions and the structure of the phase space of a shock-oscillatory system with a constant restoring force. Izv. USSR Acad. Sci. OTV, Mech. Eng. (2) (1964)
66. Bykhovsky, I.I.: On the parameters of vibratory hammers and vibratory hammers. On Sat. Pile Equipment, M., TsINTIAM (1964)
67. Bykhovsky, I.I.: Fundamentals of the theory of vibration technology. M. (1969)
68. Zaretsky, L.B.: Forced oscillations of a 6-spring vibro-impact mechanism with two impacts. Sat. Vib. Tech. M. (1966)

69. Zaretsky, J.I.B.: On self-synchronization of centrifugal vibration exciters of vibroimpact mechanism. Mach. Sci. (1) (1967)
70. Peterka, F.: Periodicke kmitani Mechanicke seustarvy s vnitrnimi RAZY, Strojncky casopis **16**(6) (1965)
71. Peterka, F.: Modelovani Kmitajich Soustav s razy on analogovam pocctaci. Strojnicky casopis. **16**(3) (1965)
72. Lukomsky, S.I.: Spring vibrohammers and their main parameters. NIIM stroydorkommunal, M. (1966)
73. Goldin, Y.M.: On the choice of the optimal scheme of the vibroimpact mechanism. Mach. Sci. (3) (1968)
74. Goldin, Y.M.: Comparative study of one-mass and two-mass vibroimpact mechanisms. On Sat. Research of Construction and Road Machines with the Use of Computer Technology. TsNIITE stroymash (1968)
75. Bespalova, L.V., Neimark, Y.I., Feigin, M.I.: Dynamical systems with shock interactions and the theory of nonlinear oscillations. Eng. MTT Mag. (1) (1966)
76. Neimark, Y.I.: Theory of vibrational immersion of sheet pile. Hydrotechnical Constr. (4) (1952)
77. Neimark, Y.I.: The theory of vibrational immersion and vibropulling. Eng. Collect. **16** (1953)
78. Blekhman, I.I.: Study of vibration driving of piles and sheet piles. Eng. Collect. **19** (1954)
79. Kushul, M.Y., Shlyakhtin, A.V.: On the theory of vibrational immersion of a metal rod in an elastoplastic medium. Izv. USSR Acad. Sci. OTN (1) (1954)
80. Blekhman, I.I., Dzhenelidze, G.Y.: Vibrational movement. M., Nauka (1964)
81. Blekhman, I.I.: Investigation of the process of vibration separation and vibration transportation. Eng. Collect. **2** (1952)
82. Lavendel, E.E.: Optimal tray movement law with given limits of acceleration during vibratory transportation in the tossing mode. Izv. USSR Acad. Sci., Mech. Eng. (6) (1964)
83. Han, Q., Zhang, Z., Wen, B.: Periodic motions of a dual–disc rotor system with rub–impact at fixed limiter. Proc. IMech E Part C: J. Mech. Eng. Sci. **222**, 1935–1946 (2008)
84. Han, W., Jin, D.P., Hu, H.Y.: Dynamics of an oblique–impact vibrating system of two degrees of freedom. J. Sound Vib. **217**, 795–822 (2004)
85. Han, W., Hu, H.Y., Jin, D.P.: Experimental study on dynamics of an oblique–impact vibrating system of two degrees of freedom. Nonlin. Dyn. **50**(3), 551–573 (2007)
86. Hassan, M.: Simulation of fluidelastic vibrations of heat exchanger tubes with loose supports. In: Proceedings ASME Pressure Vessels and Piping Division Conference, PVP2006/ICPVT–11, Vancouver, BC, Canada, p. 11 (2006)
87. Hassan, M.A., Hayder, M.: Modeling of fluidelastic vibrations of heat exchanger tubes with loose supports. Nucl. Eng. Design **238**, 2507–2520 (2008)
88. Batalova, Z.S.: On one nonlinear problem related to the theory of vibrational immersion. Izv.vysh.uch.zav., Radiophys. **6**(1) (1963)
89. Batalova, Z.S.: On the numerical study of dynamical systems with the help of computers. Izv. High. Educ. Head., Radiophys. **10**(3) (1967)
90. Batalova, S.: Study of vibration penetration of a sheet pile with frontal soil resistance. Eng. Mag., Mtt (1) (1966)
91. Asfar, K.R., Akour, S.N.: Optimization analysis of impact viscous damper for controlling self–excited vibrations. J. Vib. Cont. **11**(1), 103–120 (2005)
92. Shlyakhtin, A.V.: Periodic motions of a nonlinear autonomous system / with an internal combustion engine. Tr. IMASH AS USSR **22**(88) (1961)
93. Shlyakhtin, A.V.: Periodic motions of a piecewise linear system under the action of impulses. Proc. Semin. TMM (81) (1960)
94. Silvestrov, E.E., Shlyakhtin, A.V.: Movement of a vibrational system under the action of a time-dependent force. Proc. Semin. TMM (85–86) (1961)
95. Boyarinov, V.S., Neimark, Y.I.: On shaft vibrations in a ball bearing. Izv. USSR Acad. Sci., Mech. (3) (1965)

96. Podchufarov, B.M.: Dynamics of cyclic automation. Izv. High. uch. Manag., Mashinostr. (8) (1961)
97. Podchufarov, B.M.: Dynamics of cyclic automation with a short cycle time. Izv. High. uch. Head, Mech. Eng. (12) (1961)
98. Podchufarov, B.M.: Dynamics of cyclic automation of a machine with a movable base. Izv.vys.uch.zav., Mashinostroyeniye (6) (1963)
99. Yue–Gang, L., Zhi–hao, J., Chang–li, L., Bang–Chun,W.: Effects of nonlinear friction force on chaotic motion of rotor–bearing system at rub–impact faults. J. Northeast. Univ. (Nat. Sci.) **24**(9), 843–846 (2003)
100. Yue, Y., Xie, J.H.: Symmetry of the Poincaré map and bifurcations of a two–degree–of–freedom vibro–impact system. Zhendong Gongcheng Xuebao/J. Vib. Eng. **21**(4), 376–380 (2008) (in Chinese)
101. Ekaterina, P., Hendry, D.C., Wiercigroch, M.: Modelling of high frequency vibro-impact drilling. J. Mech. Sci. (2013). Forced Vibrations of a Spring-Dashpot Mechanism with Dry Friction and Backlash
102. Yadav, P., Balaga, S.R., Vyas, N.S.: Forced Vibrations of a Spring-Dashpot Mechanism with Dry Friction and Backlash. Nonlinear Dynamics (2020)
103. Yao, Y., Xub, J., Wiercigrochc, M.: Influence of workpiece imbalance on regenerative and frictional grinding chatters. Procedia IUTAM **22**, 146–153 (2017)
104. Farshi Homayoun Rooz, A., Hamidi, A.: Numerical analysis of factors affecting ground vibrations due to continuous impact pile driving. Int. J. Geomech. **17**(12), 04017107 (2017)
105. Hu, L., Palazzolo, A., Karkoub, M.: Suppression of lateral and torsional stick–slip vibrations of drill strings with impact and torsional dampers. J. Vib. Acoust. Copyright VC 2016 by ASME OCTOBER **138**, 051013–1 (2016)
106. Nechaev, R.F.: Dynamics of a vibro-impact crusher with a pair of self-synchronizing vibrators. Izv. USSR Acad. Sci., Mech. Eng. (5) (1963)
107. de Pater, A.D.: The approximate determination of the hunting movement of a railway vehicle by aid of the method of Krylov and Bogoljubov. Appl. Sient. Res. A **10**(3–4) (1961)
108. Pater, A.D.: Etude du mouvement de lacet d'un vehiculo de chemin de fer. Appl. Sci. Res. **6**(4) (1957)
109. Pater, A.D.: Oscillations of nonlinear mechanical systems with rigid limiters Tr. International Symposium on Nonlinear Oscillations. Application of the theory of nonlinear oscillations, vol. 3, Kyiv, ed. Academy of Sciences of the Ukrainian SSR (1963)
110. Butenin, N.V.: Elements of the Theory of Nonlinear Oscillations, L. (1962)
111. Kobrinsky, A.E., Stepanenko, Y.A.: Self-oscillating regimes in a servo hydraulic conduit with a feedback gap. On Sat. Pneumatic and Hydraulic Automatics, M. (1960)
112. Neimark, Y.I.: DAN USSR vol. 127 (1959)
113. Kolovsky, M.: Nonlinear theory of vibration protection systems. M. (1966)
114. Neimark, Y.I., Shilnikov, L.P.: On the application of the small parameter method to systems of differential equations with discontinuous right-hand sides. Izv. USSR Acad. Sci., OTN, Mech. Eng. (6) (1959)
115. Kolovsky, M.: On conditions for the existence of periodic solutions of a system of differential equations with discontinuous right-hand sides containing a small parameter. PMM **24**(4) (1960)
116. Xie, J.H., Ding, W.: Hopf-Hopf bifurcation and invariant torus T2 of a vibro–impact system. Int. J. Nonlin. Mech. **40**, 531–543 (2005)
117. Xie, J.H., Ding, W., Dowell, E.H., Virgin, L.N.: Hopf–flip bifurcation of high dimensional maps and application to vibro–impact systems. Acta. Mech. Sinica (Engl. Ser.) **21**(4), 402–410 (2005)
118. Pascal, M.: Dynamics and stability of a two–degree–of–freedom oscillator with an elastic stop. J. Comput. Nonlin. Dyn. **1**(1), 94–102 (2006)
119. Patrick, S., Jazar, G.N.: Frequency response analysis of piecewise nonlinear vibration isolator. In: Proceeding ASME 20th Biennial Conference Mechanical Vibration and Noise, Long Beach, CA, USA, vol. 1B, pp. 917–924 (2005a)

120. Patrick, S., Jazar, G.N.: Stability analysis of a piecewise nonlinear vibration isolator. In: Proceeding ASME Design Engineering Division, Orlando, FL, USA, vol. 118B(2), pp. 1115–1123 (2005b)
121. Bespalova, L.V.: The results of simulation of the problem of vibration and vibroimpact immersion. Izv. High. Educ. Manager, Radiophys. **3**(1) (1960)
122. Golovachev, A.S., (Ed.).: Studies of vibration and vibro-impact driving of piles. Transp., M. (1968)
123. Bebikhov, V.A.: Modeling on MN-8 vibrohammer. Izv. High. uch. Head
124. Althausen, E.E., Godes, Y.Y.: Electronic impact model. Mach. Sci. (3) (1966)
125. Althausen, E.V., Godes, Y.Y.: Modeling shock systems. Mach. Sci. (3) (1968)
126. Hayashi, T.: Forced Oscillations in Nonlinear Systems, Ed. I.L., M., (1957)
127. Pontryagin, L.S., Andronov, A.A., Witt, A.A.: On the statistical consideration of dynamical systems. J. Expert Theor. Phys. **3**(3) (1933)
128. Malakhov, A.N.: Fluctuations in self-oscillating systems. Nauka, M. (1968)
129. Levin, B.R.: Theoretical Foundations of Statistical Radio Engineering. Book One. Sov. radio, M. (1966)
130. Levin, B.R.: Theoretical foundations of statistical radio engineering. Book Two. Sov. radio, M. (1968)
131. Rytov, O.M.: On the theory of fluctuations in self-oscillating systems with piecewise linear characteristics. Izv. High. Academician, Radiophys. **2**(1) (1959)
132. Sadekov, R.K.: On the issue of fluctuations in piecewise linear self-oscillatory systems, Izv.vyssh.uch.zav., Radiophys. **3**(5) (1960)
133. Blekhman, I.I., Khaiman, V.Y.: On the separation of loose mixtures in oscillating vessels. III Congress of Mechanics. Abstract of reports, M., (1968)
134. Blekhman, I.I.: On the dynamics of group foundations for non-balanced machines. Inzh.zhurnal, MTT. No. 11967
135. Dimmentbert, M.F.: Amplitude-frequency response of a system with randomly changing parameters. Eng. Mag. MTT, No. 2 (1966)
136. Dimmentbert, M.F.: Oscillations of a nonlinear system with one degree of freedom under the action of a periodic force and a change in natural frequency according to a random law. Eng. Mag. MTT, No. 4 (1966)
137. Dimmenberg, M.F.: Some problems of dynamics of systems with randomly variable parameters. 3rd All-Union Congress on Theoretical and Applied Mechanics, Abstracts of Reports, M., (1968)
138. Kovaleva, A.: Response analysis of structures with motion limiters subjected to random loading. J. Struct. Control. **7**, 191–202 (2000)
139. Kovaleva, A.S.: Stochastic dynamics of flexible systems with motion limiters. Nonlin. Dyn. **36**, 313–327 (2004)
140. Masri, S.F., Ibrahim, A.M.: Response of the impact damper to stationary random excitation. J. Acoust Soc. Amer. **53**(1), 200–211 (1973)
141. Masri, S.F., Mariamy, Y.A., Anderson, J.C.: Dynamic response of a beam with a geometric nonlinearity. ASME J. Appl. Mech. **48**, 404–410 (1981)
142. Masri, S.F., Stott, S.J.: Random excitation of a nonlinear vibration neutralizer. ASME (Paper) No. 77–DET–94, September 26–30, p. 9 (1977)
143. Song, L.L., Iourtchenko, D.V.: Analysis of stochastic vibro–impact systems with inelastic impacts. Mech. Solids **41**(2), 146–154 (2006)
144. Namachchivaya, N.S., Park, J.H.: Stochastic dynamics of impact oscillators. ASME J. Appl. Mech. **72**, 862–870 (2005)
145. Wen, J.M., Feng, Q.: Two–stage stochastic model on rattling vibration with amplitude modulation. Shock. Vib. **11**(5–6), 693–702 (2004)
146. Field, R.V., Jr., Epp, D.S.: Development and calibration of a stochastic dynamics model for the design of a MEMS inertial switch. Sens. Actuators A **134**(1), 109–118 (2007)
147. Babitsky, V.I., Kolovsky, M.Z.: Oscillations of a linear system with limiters under random loads. Eng. Mag., MTT. No. 3 (1967)

148. Babitsky, V.I., Kolovsky, M.Z.: Investigation of oscillations of a linear system with limiters by exact and approximate methods. Mach. Sci. (4) (1967)
149. Goldin, Y.M., Zaretsky, L.B., Lukomsky, S.I.: Conference on the main problems of the theory of machines and mechanisms. Abstracts of Reports., M-T, (1967)
150. Goldin, Y.M., Zaretsky, L.B.: Application of mathematical modeling for the study of shock-oscillatory systems with a random recovery factor. VNShK Troydormash, M., (15) (1968)
151. Kobrinsky, A.A.: Forced motion of a vibro-impact system with a random recovery factor. Mashinovedenie (1968)
152. Kobrinsky, A.A.: Stochastic motion of a single-mass vibroimpact system. Mach. Sci. (1) (1969)
153. Weidenhammer, F.: Stabilitat sbedingungen fur Schwingungssysteme mit zufalliger Parametererregung durch weibes Rauschen, Ingen.-Archiv **35**, N1, (1–9) (1966)
154. Khasminsky, R.Z.: Stability of systems of differential equations under random perturbations of their parameters. Ed. Sci., M. (1969)
155. Nevelson, M.B., Khasminsky, R.Z.: Problems of information transmission. **2**(3) (1966)
156. Nevelson, M.B.: On the operation of a linear system with small random excitations of its parameters. PMM **31**(3) (1967)
157. Nevelson, M.B., Khasminsky, R.: On the stability of a linear system under random excitations of its parameters. PMM (2) (1966)
158. Nevelson, M.B.: Some remarks on the stability of a linear stochastic system. PMM (6) (1966)
159. Khasminsky, R.Z.: On stability in the first approximation of stochastic systems. PMM **3**(6) (1967)
160. Katz, I.Y.: On stability in first approximation of a system with random delay. PMM **31**(3) (1967)
161. Brusin, V.A., Tai M.J.I.: Absolute stochastic stability. Izv. High. uch. Head, Radiophys. **10**(7) (1967)
162. Kolomeets, V.G., Korenvsky, D.G.: On the stability of linear systems with random excitations. Appl. Mech. **3**(8)
163. Kats, I.Y., Krasovsky, N.N.: On the stability of systems with random parameters. PMM **27**(5) (1960)
164. Bespalova, L.V., Metrikin, V.S.: On the theory of a two-mass model of a vibrohammer. Izv. High. uch. Head, Radiophys. **9**(2) (1966)
165. Bespalova, L.V., Metrikin, V.S.: Some questions of the theory of a two-mass model of a vibrohammer. Proceedings of the Interuniversity Scientific Conference on Electric Percussion Machines. Novosibirsk (1967)
166. Bespalova, L.V., Metrikin, V.S.: On the theory of a two-mass model of a vibrohammer. All-Union Interuniversity Symposium on Applied Mathematics and Cybernetics. Abstracts of Reports. Gorky (1967)
167. Metrikin, V.S.: On the stability of one model of vibrohammer. Izv. High. uch. Head, Radiophys. 11(3) (1968)
168. Bespalova, L.V., Lobanov, V.I., Metrikin, V.S.: The results of the calculation on the computer of the boundaries of the stability of the movement of the vibro-impact compactor of the slope of the prism. Bull. All-Union Res. Inst. Railway. dor. Transp. (1) (1969)
169. Bespalova, L.V., Metrikin, V.S.: On the stability of a two-mass model of a vibrohammer. Mechanics of Machines, USSR Academy of Sciences, ed. Science, M., (1969)
170. Bespalova, L.V., Metrikin, V.S.: Influence of viscous friction on the stability of a vibrohammer. Izv. USSR Acad. Sci., MTT, (2) (1969)
171. Metrikin, V.S.: On the theory of a two-mass model of a vibrorammer. Izv. High. uch. Head, Radiophys. (3) (1969)
172. Metrikin, V.S.: On the theory of vibrohammer with randomly changing parameters. Izv. High. uch. Head, Radiophys. (4) (1970)

Chapter 2
Dynamics of Two-Mass Vibro-impact Mechanisms with a Fixed Limiter

Abstract The first chapter examines the nonlinear dynamics of various two-mass models of a vibration impactor with a fixed limiter. The consideration begins with a general two-mass model, consisting of two masses (impact mass) and (vibrator), fixed by means of springs and interconnected by a connecting spring. An external harmonic force acts on the upper mass. The amplitude of oscillations of the lower mass is limited by a stationary barrier (limiter), upon contact with which an instantaneous, generally speaking, elastic impact occurs, characterized by a coefficient of recovery. In § 2.1, the problem of studying forced oscillations of a vibro-impact system is reduced to the study of a point mapping of a four-dimensional hyperplane (corresponding to the limiter plane) into itself in a five-dimensional phase space. The structure of phase space is considered, and possible types of impact motions are indicated in the case of both elastic and inelastic impacts. It is shown that any periodic movement can be characterized by two numbers m and n, where m is the number of blows during the period of movement, and n is the multiplicity of the period of movement to the period of the external force. In § 2.2, in the parameter space, the coordinates of fixed points corresponding to the periodic single-impact multiple mode of motion are found in parametric form. The boundaries of the existence of these periodic movements are determined. Based on the theory of point mappings and the theory of bifurcations, equations of bifurcation surfaces are given that correspond to the violation of the conditions of existence and stability of fixed points (N-surfaces) and the exit of a closed phase trajectory beyond the domain of definition of the point transformation under study (C-surfaces). Bifurcation C-surfaces, separating single-impact periodic movements from multi-impact ones, correspond to the appearance of additional impacts during the period of movement. It is noted that for parameter values at which the gap is greater than the amplitude of forced oscillations, along with shockless and quasi-periodic movements, movements with a rigid nature of occurrence are possible. In § 2.3, the dynamics of a spring-loaded two-mass vibratory impactor is studied. It is shown that the regions of existence of single-impact multiple periodic modes of motion, previously found from the condition of reality and positivity of the impact velocity, are significantly narrowed due to the loss of their stability. To partition the parameter space into regions of stability, a

V. Metrikin and I. Leonid, *Forced Oscillations of Multidimensional Highly Nonlinear Systems*, Advanced Structured Materials 222,
https://doi.org/10.1007/978-981-97-6779-3_2

characteristic polynomial of the fourth degree was obtained, based on which bifurcation N-surfaces for different parameter values. Of the stability regions, the most significant are those located near the resonant frequencies of the corresponding linear system without a limiter. A study of the location of bifurcation C-surfaces showed that they cut off a significant part from the areas of N-stability. A methodology for constructing both N-stability regions and bifurcation C-surfaces is presented. In § 2.4, a model of a two-mass non-spring vibratory hammer was studied. The stability of single-impact multiple periodic movements was studied. The boundaries of the stability regions of these movements are found, and bifurcation C-surfaces are constructed. In § 2.5, stable single-impact multiple periodic movements for a two-mass vibration-impact unspring compactor are studied and found. For an absolutely elastic impact ($R = 1$), some general properties of bifurcation N-surfaces are given, and a method for constructing bifurcation surfaces N_φ is also indicated. The location of stability regions in three-dimensional parameter space has been studied.

Keywords Mathematical model · Poincaré point mapping method · Nonlinear dynamics · Stability · Bifurcation · Vibration-impact mechanism · Characteristic number · Characteristic polynomial

2.1 Problem Formulation

Currently, many models of vibro-impact systems are well known [1, 2]. All of them can be divided into the following classes:

(a) vibro-strikers with directed exiting force,
(b) vibro-strikers with non-directed exiting force,
(c) vibro-strikers with special impact mass,
(d) vibro-strikers without special impact mass.

In the simplest model of a vibro-striker [3], the vibrator and the striker represent a single-mass striking against a fixed limiter. Due to the relatively high rate of the system work, the vibrator experiences frequent impacts against the limiter that sharply decreases the service life of the striker. Therefore, a number of other designs of vibro-impact systems were proposed in which the vibrator and striker are separated. In the proposed designs, the vibrator itself does not experience impacts against the fixed limiter that significantly increases the service life of such systems. However, the introduction of a special impact mass increases the number of essential system parameters that complicates the adjustment of the system to a required mode. When in service, such systems are required to receive maximum pulses upon impact. This condition may be satisfied only with optimal tuning of the system parameters which values can be specified after a detailed theoretical study of the considered vibro-impact systems (in particular, allocation in the parameter space the existence and stability regions for various periodic motions). It is also necessary to provide

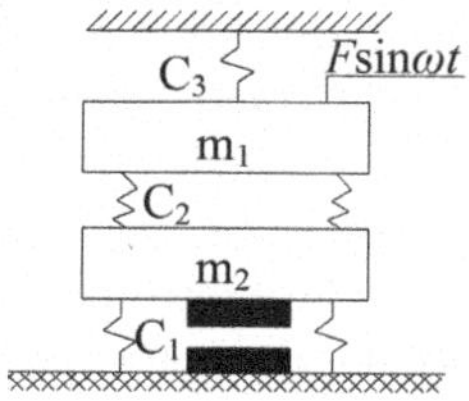

Fig. 2.1 Physical diagram of a general two-mass vibration-impact mechanism

theoretical study because the partition of the parameter space into stability regions is a rather thin structure which is essential for selecting the optimal parameter values.

Below, we consider a vibro-impact system with a special impact mass, the design scheme of which is shown in Fig. 2.1.

It consists of two masses (impact mass) and m_2 (vibrator), fixed by means of springs C_1 and C_3, interconnected by a connecting spring C_2. An external periodic force $F \sin \omega t$ acts on the mass m_2. The motion of mass m_1 is limited from below by a fixed barrier, in contact with which an inelastic impact is generated with a speedy recovery coefficient $0 \leq R \leq 1$.

For mathematical description of the system, we make a number of assumptions (A):

1. energy is dissipated only by impacts, and there is no loss of energy between impacts,
2. the springs work in the linear part of their characteristics with an elastic coefficient,
   - at the upper springs—C_3,
   - at the connecting springs—C_2,
   - at the lower springs—C_1.
3. the impact occurs instantly with the coefficient of restitution R varying within $0 \leq R \leq 1$,
4. the displacement of the limiter is not considered,
5. there is no collision between the masses.

Further, for studying various vibro-impact systems, we will follow these assumptions.

The reference point is the position of the system in which all springs are in an undeformed state. Then, under the assumptions made, the equations of motion of the system in the intervals between impacts are written in the form

$$\begin{aligned} m_1\ddot{x}_1 + (C_1 + C_2)x_1 - C_2x_2 &= \mathrm{P}_1, \\ m_2\ddot{x}_2 + (C_2 + C_3)x_2 - C_2x_1 &= \mathrm{P}_1 + F\sin\omega t. \end{aligned} \tag{2.1.1}$$

where P_1 is the weight of the impact mass, P_2 is the constant force acting on the mass m_2,x_1 and x_2 are the displacements of the lower and upper masses, the positive direction of which is indicated by the arrow in Fig. 2.1. The gap between the striker and limiter is denoted by x_0. It can be either positive or negative, depending on whether there is a gap between the striker and the oscillation limiter in the initial state, or whether the striker is pressed against the limiter, i.e., there is pre-compression of springs.

At the moment of the striker contact with the limiter at $x_1 = x_0$, an inelastic impact occurs, characterized by the ratio

$$\dot{x}_1^+ = -R\dot{x}_0^-, \tag{2.1.2}$$

where $\dot{x}_0^-$ and $\dot{x}_1^+$ are velocities right before and right after impact.

Thus, the general two-mass model is substantially nonlinear, non-autonomous system with two degrees of freedom.

The system of Eqs. (2.1.1) and (2.1.2) in dimensionless variables

$$\xi_1 = \frac{m_1\omega^2}{F}\left\{x_1 - \frac{C_2 + C_3}{C_1C_2 + C_1C_3 + C_2C_3}\left[\mathrm{P}_1 + \mathrm{P}_2\frac{C_1 + C_2}{C_2}\right] + \frac{\mathrm{P}_2}{C_2}\right\}, \tag{2.1.3}$$

$$\xi_2 = \frac{m_2\omega^2}{F}\left\{x_2 - \frac{C_2}{C_1C_2 + C_1C_3 + C_2C_3}\left[\mathrm{P}_1 + \mathrm{P}_2\frac{C_1 + C_2}{C_2}\right]\right\}, \tag{2.1.4}$$

will be rewritten in the form at $\xi_1 < \xi_0$,

$$\begin{aligned} &\ddot{\xi}_1 + \lambda_2^2(\phi\xi_1 - \xi_2) + \lambda_1^2\xi_1 = 0, \\ &\ddot{\xi}_2 + \lambda_2^2(\xi_2 - \phi\xi_1) + \mu\lambda_2^2\xi_2 = \sin\tau. \end{aligned} \tag{2.1.5}$$

at $\xi_1 = \xi_2$,

$$\dot{\xi}_1 = -R\dot{\xi}_0. \tag{2.1.6}$$

where the following dimensionless parameters are introduced

$$\lambda_1^2 = \frac{C_1}{m_1\omega^2}, \lambda_2^2 = \frac{C_2}{m_2\omega^2}, \phi = \frac{m_2}{m_1}, \mu = \frac{C_3}{C_2}. \tag{2.1.7}$$

It is seen from (2.1.5) and (2.1.6) that the system is characterized by six parameters $\lambda_1, \lambda_2, \phi, \xi_0, R, \mu$ changing within limits

$$0 \le R \le 1, 0 \le \lambda_1, \lambda_2, \phi, \mu < +\infty, -\infty < \xi_0 < +\infty. \tag{2.1.8}$$

The phase space of the considered general two-mass system with two degrees of freedom is five-dimensional in coordinates $\xi_1, \xi_2, \dot{\xi}_1, \dot{\xi}_2, \tau$. Hyperplane $\xi_1 = \xi_0$ (denoted by Π) corresponds in the phase space to a limiter plane. In accordance with Eqs. (2.1.5), (2.1.6), the phase space of the system is divided by the hyperplane Π into two parts $\xi_1 < \xi_0$ (we denote this part of the half-space by G) and $\xi_1 > \xi_0$ (denoted by W). All possible motions of the system lie in $\overline{G}$ region that includes the half-space G and the hyperplane P. For impact motions of the system at $R \neq 0$, the image point moves to $\overline{G}$ in the following way: moving in the region G, it hits the hyperplane Π at a point $M_0(\tau_0, \dot{\xi}_1^0, \xi_2^0, \dot{\xi}_2^0)$ and then instantly moves along the hyperplane Π to a point $M_{0'}(\tau_0, -R\dot{\xi}_1^0, \xi_2^0, \dot{\xi}_2^0)$. If $\dot{\xi}_1^0 \neq 0$, then the imaging point, having left the hyperplane Π, moves according to Eq. (2.1.5) in the domain G until it falls again onto the hyperplane Π at a point $M_1(\tau_1, \dot{\xi}_1, \xi_2, \dot{\xi}_2)$. In addition to the motions described above, the motion when the image point, moving from region G to the hyperplane Π, reaches a point $M_0(\tau_0, 0, \xi_2^0, \dot{\xi}_2^0)$ that corresponds to the striker touching the plane of the limiter, is also possible at $R \neq 0$. At this moment $\tau = \tau_0$, its acceleration is greater than zero $\ddot{\xi}_1^0 > 0$, then the image point will move along the hyperplane with a different trajectory than in the case $\dot{\xi}_1^0 \neq 0$ until $\ddot{\xi}_1$ becomes less than zero. This part of the trajectory corresponds to a prolonged stop of a striker, which is pressed against the oscillation limiter. Thus, in the case of impact motions of the system, the image point will each time hit the hyperplane Π at the moment of impact and, therefore, the study of its motion can be reduced to studying the mapping of the hyperplane Π into itself.

In the case of absolutely inelastic impact ($R = 0$), the imaging point, having fallen from the region G onto the point M_0, instantly moves to the point $M_{0'}(\tau_0, 0, \xi_2^0, \dot{\xi}_2^0)$ and then, if at the moment $\tau = \tau_0$ acceleration $\ddot{\xi}_1^0 < 0$, it falls on to the G region, then the imaging point moves along the hyperplane Π until the acceleration $\ddot{\xi}_1$ becomes zero, after which it again moves to the region G.

In the phase space, a closed trajectory corresponds to periodic motion of the system. From differential Eq. (2.1.5), as in the case of the single-mass vibro-striker model [3], it follows that periodic motions of the system can only have the period that is a multiple of the driving force period, i.e., $T = 2\pi n$, where $n = 1, 2, 3, \ldots$. These motions are called n-fold. For the period of motion, the image point can fall several times onto hyperplane Π ($m = 1, 2, 3, \ldots$) that corresponds to the motion when impact mass impacts m times. Such motions are called without-, -single-, two-, three-, etc., impact.

Thus, as in the case of a single-mass vibro-striker, the periodic motions of a two-mass system can be characterized by two numbers m and n.

By integrating the system (2.1.5) with the initial conditions taken on the hyperplane Π $\xi_1 = \xi_0, \xi_2 = \xi_2^\circ, \dot{\xi}_1 = -R\dot{\xi}_0, \dot{\xi}_2 = \dot{\xi}_2^\circ$ for $\tau = \tau_0$, the solution can be written in the following form

$$\xi_1 = a \sin \tau + \sum_{i=1}^{2} k_i \left[B_i \cos \eta_i(\tau_i - \tau_0) + A_i \eta_i^{-1} \sin \eta_i(\tau - \tau_0)\right],$$

$$\dot{\xi}_1 = a \cos \tau + \sum_{i=1}^{2} k_i [A_i \cos \eta_i(\tau - \tau_0) - B_i \eta_i \sin \eta_i(\tau - \tau_0)],$$

$$\xi_2 = b \sin \tau + \sum_{i=1}^{2} \left[B_i \cos \eta_i(\tau - \tau_0) + A_i \eta_i^{-1} \sin \eta_i(\tau - \tau_0)\right],$$

$$\dot{\xi}_2 = b\cos\tau + \sum_{i=1}^{2}[A_i\cos\eta_i(\tau-\tau_0) - B_i\eta_i\sin\eta_i(\tau-\tau_0)]. \tag{2.1.9}$$

where the following notation is introduced

$$\begin{aligned} A_1 &= (k_2-k_1)^{-1}\left[R\dot{\xi}_0 + k_2\dot{\xi}_2^\circ + (a-bk_2)\cos\tau_0\right] \\ B_1 &= (k_2-k_1)^{-1}\left[-\xi_0 + k_2\xi_2^\circ + (a-bk_2)\sin\tau_0\right]. \end{aligned} \tag{2.1.10}$$

A_2 and B_2 are derived from A_1 and B_1 by replacing

$$k_2 = \frac{(1+\mu)\lambda_2^2 - \eta_2^2}{\phi\lambda_2^2} \quad \text{and} \quad k_1 = \frac{(1+\mu)\lambda_2^2 - \eta_1^2}{\phi\lambda_2^2}. \tag{2.1.11}$$

with k_1 and k_2, respectively, where η_1 and η_2 are roots of frequency equations

$$\eta^4 - \left[(1+\phi+\mu)\lambda_2^2 + \lambda_1^2\right]\eta^2 + \lambda_1^2\lambda_2^2 + \mu\left(\lambda_1^2 - 1 + \phi\lambda_2^2\right)\lambda_2^2 = 0. \tag{2.1.12}$$

and

$$a = \lambda_2^2\,\Delta^{-1}, b = \left(\lambda_1^2 + \lambda_2^2\phi - 1\right)\Delta^{-1}. \tag{2.1.13}$$

are the amplitudes of the forced oscillations of the impact mass and vibrator, respectively, where

$$\Delta = \left(1-\eta_1^2\right)\left(1-\eta_2^2\right). \tag{2.1.14}$$

Further, we agree to take a big root for η_1, herewith the following inequality will be held

$$k_2 - k_1 > 0. \tag{2.1.15}$$

The moment of the next impact is found from the first equation of system (2.1.9) by assuming $\xi_1 = \xi_0$. Three other equations of this system determine the position and velocity of the vibrator at the moment of impact and the impact mass velocity. Thus, transformation of the hyperplane Π into itself (we call it the S-transformation) is determined by the following system of four equations.

$$\xi_0 = a\sin\tau_1 + \sum_{i=1}^{2}k_i\left[B_i\cos\eta_i(\tau_1-\tau_0) + A_i\eta_i^{-1}\sin\eta_i(\tau_1-\tau_0)\right],$$

$$\dot{\xi} = a\cos\tau_1 + \sum_{i=1}^{2}k_i[A_i\cos\eta_i(\tau_1-\tau_0) - B_i\eta_i\sin\eta_i(\tau_1-\tau_0)],$$

$$\xi_2 = b \sin \tau_1 + \sum_{i=1}^{2} \left[B_i \cos \eta_i (\tau_1 - \tau_0) + A_i \eta_i^{-1} \sin \eta_i (\tau_1 - \tau_0) \right],$$

$$\dot{\xi}_2 = b \cos \tau_1 + \sum_{i=1}^{2} [A_i \cos \eta_i (\tau_1 - \tau_0) - B_i \eta_i \sin \eta_i (\tau_1 - \tau_0)]. \tag{2.1.16}$$

It should be noted that from all the roots $\tau_1^i (i = 1, 2, \ldots, n)$ of the first Eq. (2.1.16), which determines the moment of the next impact $\tau_1 = \tau_1^k$, followed after τ_0, it is necessary to choose the smallest one that satisfies the following condition $\tau_1^k = \min\left(\tau_1^1, \tau_1^2, \ldots, \tau_1^n\right)$, $\tau_1 > \tau_0$, $1 \leq k \leq n$.

2.2 Finding Stable Periodic Motions of a Two-Mass System

Finding periodic motion is reduced to finding a fixed point of the point mapping Sat unidirectional motion and S^m-mapping (m times of the repeated S mapping) at m impact motion.

Writing down formulas S^m-mappings and using periodicity conditions,

$$\dot{\xi}_1^m = \dot{\xi}_0, \dot{\xi}_2^m = \dot{\xi}_2^\circ, \xi_2^m = \xi_2^\circ, \tau_m - \tau_0 = 2\pi n (n, m = 1, 2, \ldots). \tag{2.2.1}$$

we will have the following system of equations $4(m+1)$ to determine the coordinates of the invariant point of the S^m point mapping

$$\xi_0 = a \sin \tau_e + \sum_{i=1}^{2} k_i \left[B_i \cos \eta_i (\tau_e - \tau_{e-1}) + A_i \eta_i^{-1} \sin \eta_i (\tau_e - \tau_{e-1}) \right],$$

$$\dot{\xi}_1^e = a \cos \tau_e + \sum_{i=1}^{2} k_i \left[A_i \cos \eta_i (\tau_e - \tau_{e-1}) - B_i \eta_i \sin \eta_i (\tau_e - \tau_{e-1}) \right],$$

$$\xi_2^e = b \sin \tau_e + \sum_{i=1}^{2} \left[B_i \cos \eta_i (\tau_e - \tau_{e-1}) + A_i \eta_i^{-1} \sin \eta_i (\tau_e - \tau_{e-1}) \right],$$

$$\dot{\xi}_2^e = b \cos \tau_e + \sum_{i=1}^{2} \left[A_i \cos \eta_i (\tau_e - \tau_{e-1}) - B_i \eta_i \sin \eta_i (\tau_e - \tau_{e-1}) \right],$$

$$\xi_2^m = \xi_2^0, \dot{\xi}_1^m = \dot{\xi}_1^0, \dot{\xi}_2^m = \dot{\xi}_2^0, \tau_m - \tau_0 = 2\pi n (e = 1, 2, \ldots, m). \tag{2.2.2}$$

Solving the system (2.2.2) for any integer n for single-impact periodic motions $m = 1$, we obtain the coordinates of the fixed point in parametric form

$$\sin \tau_0 = \xi_0 a^{-1} - [2af]^{-1} F \dot{\xi}_0, \cos \tau_0 = b^{-1} \dot{\xi}_2^0,$$
$$\xi_2^0 = ba^{-1} \xi_0 + [2af]^{-1} (1 + R)$$

$$\left[\eta_2(a - bk_1)ctg(\pi n\eta_1) - \eta_1(a - bk_2)ctg(\pi n\eta_2)\right],$$
$$\dot{\xi}_2^0 = (1 - R)(2a)^{-1}b\dot{\xi}_0. \tag{2.2.3}$$

Impact velocity $\dot{\xi}_0$ is determined as the solution of the equation

$$[F^2 + (1 - R)^2f^2]\dot{\xi}_0^2 - 4fF\xi_0\dot{\xi}_0 + 4f^2(1 - a^2)\xi_0^2 = 0. \tag{2.2.4}$$

where f and F are denoted.

$$f = (k_2 - k_1)\eta_1\eta_2,$$
$$F = (1 + R)\left[k_1\eta_2 ctg(\pi n\eta_1) - k_2\eta_1 ctg(\pi n\eta_2)\right]. \tag{2.2.5}$$

Solving the Eq. (2.2.4) relatively $\dot{\xi}_0$, we obtain.[1]

$$\dot{\xi}_0 = \frac{2f}{F^2 + (1 - R)^2f^2}$$
$$\left\{F\xi_0 \pm \left[a^2(F^2 + (1 - R)^2f^2) - (1 - R)^2f^2\xi_0^2\right]^{\frac{1}{2}}\right\}. \tag{2.2.6}$$

Since impact periodic motions can occur only when the impact velocity $\dot{\xi}_0$ has a real and positive value, the boundaries of existence and periodic motion are derived from (2.2.6) as follows: if $F\xi_0 < 0$, then for positivity $\dot{\xi}_0$ it is necessary to satisfy the inequality $|\xi_0F| \le \left[a^2F^2 + (1 - R)^2f^2(a^2 - \xi_0^2)\right]^{\frac{1}{2}}$, if $F\xi_0 > 0$, then the condition for the positivity of the impact velocity coincides with the boundaries of reality $a^2F^2 + (1 - R)^2f^2(a^2 - \xi_0^2) \ge 0$ of single-impact periodic motions, which in the parameter space of the system can be written as

$$|\xi_0| = |a|, (F\xi_0 < 0),$$
$$|\xi_0| = |a|\left[\frac{F^2}{(1 - R)^2f^2} + 1\right]^{\frac{1}{2}}, (F\xi_0 > 0). \tag{2.2.7}$$

$$n\eta_1 = e, n\eta_2 = k, (n \ne k, n \ne l, k, n, l = 1, 2, 3, \ldots). \tag{2.2.8}$$

Conditions (2.2.7) and (2.2.8) determine the regions in the parameter space where single-impact n-multiple periodic motions are possible. In the regions defined by the boundaries of the existence of fixed points of a point mapping, it is necessary to single out those values of the parameters of the system at which the periodic motions that correspond to the obtained fixed points (2.2.7) are stable.

To study stability in a small periodic motion, a characteristic polynomial $\chi(z) = 0$ is derived, as a result of point mapping linearization in a neighborhood of a fixed point. Let $M*\left(\tau*, \dot{\xi}_1^*, \xi_2^*, \dot{\xi}_2^*\right)$ be a fixed-point S-transformation and $A_0\left(\tau_0, \dot{\xi}_1^0, \xi_2^0, \dot{\xi}_2^0\right)$ a

[1] Equations (2.2.4) coincide (up to notation) with the corresponding equations in [4], and for $\phi = 0$ with similar relations in works [5–9].

point close to $M*$. After applying the S mapping, the coordinates of the point $M*$ remain unchanged, and the coordinates of A_0 point are changed and A_0 point is converted into $A_1\left(\tau_1, \dot{\xi}_1^1, \xi_2^1, \dot{\xi}_2^1\right)$ point. We find the relation between the deviations of $\triangle\ \tau_0$, $\triangle\ \dot{\xi}_1^0$, $\triangle\ \xi_2^0$, $\triangle\ \dot{\xi}_2^0$ and $\triangle\ \tau_1$, $\triangle\ \dot{\xi}_1^1$, $\triangle\ \xi_2^1$, $\triangle\ \dot{\xi}_2^1$ for coordinates of the points A_0 and A_1 and the coordinates of the fixed point A_1 by restricting ourselves to linear terms. Linearizing the equations of the point mapping (2.1.16) in a neighborhood of a fixed point $M*$, we find

$$\triangle\ x_k^1 = \sum_{j=1}^{4} a_{kj}\ \triangle\ x_k^0,\ (k = 1, 2, 3, 4). \tag{2.2.9}$$

where $\triangle\ x_k$ denotes $\triangle\ \tau$, $\triangle\ \dot{\xi}_1$, $\triangle\ \xi_2$, $\triangle\ \dot{\xi}_2$, respectively. Coefficients a_{kj} in Eq. (2.2.9) are determined by the following

$$a_{kj} = -\left(\frac{\partial F_k/\partial x_j^0}{\partial F_k/\partial x_k^1}\right)_{M*},\ (k, j = 1, 2, 3, 4). \tag{2.2.10}$$

where x_1, x_2, x_3, x_4 denote $\tau, \dot{\xi}_1, \xi_2, \dot{\xi}_2$, respectively, and F_1, F_2, F_3, F_4 are the essence of the expression, obtained by transferring all the terms in Eq. (2.1.16) to the left. Assuming that the relationship between the deviations $\triangle\ x_k^1$ and $\triangle\ x_k^0$ is linear, we put in (2.2.9) $\triangle\ x_k^1 = z\ \triangle\ x_k^0$, then we obtain

$$\sum_{j=1}^{4}\left(a_{kj} - z\right)\ \triangle\ x_k^0 = 0,\ (k = 1, 2, 3, 4). \tag{2.2.11}$$

The system (2.2.11) is a system of homogeneous algebraic equations with respect to $\triangle\ x_k^0$, and we will have a nontrivial solution if and only if its determinant is zero [10]. Therefore, we have

$$\det\left\|a_{kj} - z\right\| = 0,\ (k, j = 1, 2, 3, 4). \tag{2.2.12}$$

Expanding the determinant (2.2.12), we obtain a characteristic polynomial of the fourth degree with respect to z, which determines the stability of a fixed point of point mapping. For stability of a fixed point, it is necessary that all the roots of the characteristic polynomial (2.2.12) lie inside the unit circle $|z| \leq 1$ and it is sufficient that they lie strictly inside the unit circle $|z| < 1$.

It was shown in [11–13] that with a continuous change in the parameters, the stability of periodic motion is violated only on three surfaces, and the equations of which are obtained from characteristic Eq. (2.2.12) on substituting $z = +1$, $z = -1$, $z = e^{\pm i\varphi}$, $0 \leq \varphi \leq \pi$ into it. These surfaces were named N_+, N_-, N_φ, respectively. In what follows, we assume that if the stability region of periodic motions is limited by surfaces, then this boundary is N-stability.

After constructing the boundary of N-stability, we obtain N-stability regions of periodic motions of $(n, m = 1)$ type. However, motions with $(n, m > 1)$ values can also be found in the selected regions of N-stability, since when we find the coordinates of a fixed point, the condition $\xi_1 < \xi_0$ for all $\tau_0 < \tau < \tau_1(\tau_1 - \tau_0 = 2\pi n)$ is not used. The contact of the motion trajectory of the impact mass with the limiter plane is a boundary case for this condition to be met. Therefore, the outcome of the phase trajectory from the definition domain of the point mapping is related to the contact of the motion trajectory of the lower mass at the moment $\tau = \tau*$ on the limiter surface, where $\tau_0 < \tau* < \tau_1$. The emerging of trajectory contact at the moment of $\tau*$ means that with a further change in system parameters, an additional impact occurs for the motion period. The values of the parameters corresponding to an additional impact during the period of motion are bifurcational. The whole set of the bifurcation parameter values, which are being in their turn the points of discontinuity for point mapping, represent a surface in the system parameter space, denoted by C_τ [14].

From the foregoing, it follows that the region of existence and stability of the periodic motions under study are limited by N- and C-surfaces in order to determine the latter we have the conditions:

$$\xi_1(\tau*) = \xi_0, \dot{\xi}_1(\tau*) = 0, \text{at} \quad \tau_0 < \tau* < \tau_1(\tau_1 - \tau_0 = 2\pi n). \tag{2.2.13}$$

Or in an expanded form the conditions, (2.2.13) can be written in the form

$$\xi_0 = a\sin(\tau_0 + \beta) + \sum_{i=1}^{2} k_i\left[B_j\cos\eta_i\beta + A_i\eta_i^{-1}\sin\eta_i\beta\right],$$

$$0 = a\cos(\tau_0 + \beta) + \sum_{i=1}^{2} k_i\left[A_j\cos\eta_i\beta - B_i\eta_i\sin\eta_i\beta\right]. \tag{2.2.14}$$

where $\beta = \tau * - \tau_0$ is denoted by β.

Substituting the coordinates of the fixed point (2.2.3) into (2.2.14) and by excluding ξ_0 from the equation, we obtain

$$\Lambda_1\Lambda_4 - \Lambda_2\Lambda_3 = 0. \tag{2.2.15}$$

where the following notions are introduced

$$\Lambda_1 = f_7g_1 + f_8g_4 + f_9g_2 + f_{10}g_6 + f_{16},$$
$$\Lambda_2 = f_8g_3 + f_{10}g_5 - \alpha + f_{15},$$
$$\Lambda_3 = f_{11}g_1 + f_{12}g_4 + f_{13}g_2 + f_{14}g_6 + f_{17},$$
$$\Lambda_4 = f_{12}g_3 + f_{14}g_5 + f_{18},$$
$$g_1 = \left[R + k_2f_5 + (a - bk_2)f_6\right], g_2 = \left[R + k_1f_5 + (a - bk_1)f_6\right],$$
$$g_3 = \left[k_2f_3 + (a - bk_2)f_1 - 1\right], g_4 = \left[k_2f_4 + (a - bk_2)f_2\right],$$

$$
\begin{aligned}
g_5 &= \left[k_1 f_3 + (a - bk_1) f_2 - 1\right], g_6 = \left[k_1 f_4 + (a - bk_1) f_2\right]. \\
f_1 &= a^{-1}, f_2 = -F(2af)^{-1}, f_3 = ba^{-1}, \\
f_4 &= \frac{1+R}{2aR}\left[\eta_2 (a - bk_1) ctg \pi n \eta_1 - \eta_1 (a - bk_2) ctg \pi n \eta_2\right], \\
f_5 &= \frac{1-R}{2a} b, f_6 = \frac{1-R}{2a}, f_7 = k_1 \eta_1^{-1} \sin \eta_1 \beta, f_8 = k_1 \cos \eta_1 \beta, \\
f_9 &= -k_2 \eta_2^{-1} \sin \eta_2 \beta, f_{10} = -k_2 \cos \eta_2 \beta, f_{11} = k_1 \cos \eta_1 \beta, f_{12} \\
&= -k_1 \eta_1 \sin \eta_1 \beta, \\
f_{13} &= -k_2 \cos \eta_2 \beta, f_{14} = k_2 \eta_2 \sin \eta_2 \beta, f_{15} = \alpha a f_1 \cos \beta, \\
\alpha &= (k_2 - k_1), \\
f_{16} &= \alpha a (f_6 \sin \beta + f_2 \cos \beta), f_{17} = \alpha a (f_6 \cos \beta - f_2 \sin \beta), \\
f_{18} &= -\alpha a f_1 \sin \beta.
\end{aligned} \tag{2.2.16}
$$

To determine the gap ξ_0 we can derive from (2.2.14)

$$
\xi_0 = \frac{\Lambda_3^2 \psi_2 l^2}{(\Lambda_4 + \psi_1 \Lambda_3)^2 + l^2 \Lambda_3^2 \psi_3}. \tag{2.2.17}
$$

where

$$
\begin{aligned}
\psi_1 &= \frac{2fF}{F^2 + (1-R)^2 f^2}, l = \psi_1 F^{-1}, \\
\psi_2 &= \left[a^2 F^2 + a^2 (1-R)^2 f^2\right], \psi_3 = (1-R)^2 f^2.
\end{aligned} \tag{2.2.18}
$$

in domain

$$
\xi_0 > \frac{\lambda_2^2}{\left|1 - \left[(1 + \phi + \mu)\lambda_2^2 + \lambda_1^1\right] + \lambda_1^2 \lambda_2^2 + \mu \lambda_2^2 (\phi \lambda_2^2 + \lambda_1^2 - 1)\right|}. \tag{2.2.19}
$$

under some initial conditions, the impact less motions of the system are possible, since the condition (2.2.19) means that the gap is larger by the amplitude value of the forced oscillations of the striker corresponding to a linear system without a limiter. It is known [3] that, in addition to the impact less motions in the region determined by (2.2.19), impact motions are also possible under certain initial conditions. Thus, both impact less and impact motions of the system can exist in the region $\xi_0 > |a|$, depending on the initial conditions. This parameter region [15] is called the region of hard excitation of oscillations.

2.3 Studying the Stability of a Single-Impact *n*-fold Periodic Motions of a Spring-Loaded Two-Mass Vibro-striker ($C_3 = 0$)

Two-mass vibro-impact systems, rigidly connected only with the processed material, are often used for pulling or driving piles, sheet piles, structures, etc. [1, 2, 16–24]. One of these models can be obtained from a common two-mass system (see Fig. 2.1), assuming $C_3 = 0$ (i.e., a system without upper springs is obtained) (Fig. 2.2).

A similar model was considered firstly by D. D. Barkhan and O. Ya. Shekhter in [25], where only the existence of periodic regimes was studied. However, as discussed below, the study of the existence of periodic motions without examining their stability does not reveal the parameters of the system for which periodic motions do exist. This is explained by the fact that in nonlinear systems in general and in systems with impact interactions in particular, the region of stability of periodic regimes is less [14, 26–34] than the existence region.

The considered model of vibro-striker consists of two masses, one of which is the impact mass (m_1), and the other is a vibrator that creates an external periodic force $F \sin \omega t$. The masses are interconnected by springs with a coefficient of elasticity C_2. The impact mass is also connected to the limiting plane by means of springs with a coefficient of elasticity C_1.

Equations of motion between impacts ($x_1 < x_0$) in dimensionless variables ξ_1, ξ_2, τ are derived from (2.1.5), (2.1.6) assuming $\mu = 0$ and have the form at $\xi_1 < \xi_0$

$$\begin{gathered}\ddot{\xi}_1 + \lambda_2^2(\phi\xi_1 - \xi_2) + \lambda_1^2\xi_1 = 0,\\ \ddot{\xi}_2 + \lambda_2^2(\xi_2 - \phi\xi_1) = \sin \tau.\end{gathered} \tag{2.3.1}$$

at $\xi_1 = \xi_0$

$$\dot{\xi}_1 = -R\dot{\xi}_0. \tag{2.3.2}$$

The behavior of the system is characterized by five parameters $\lambda_1, \lambda_2, \phi, R, \xi_0$ which are changing within the limits.

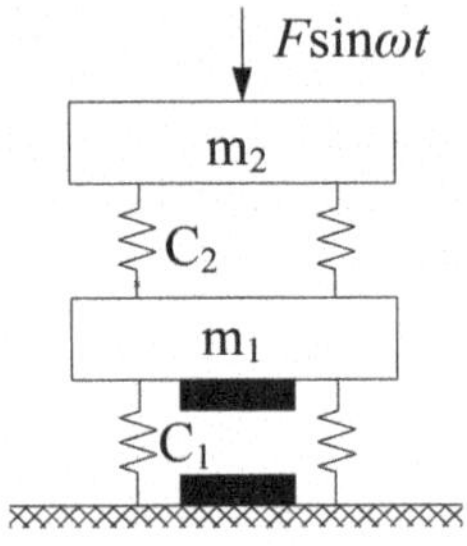

Fig. 2.2 Scheme of a two-mass vibration-impact mechanism with a special impact mass

$0 \le \lambda_1, \lambda_2, \phi \le +\infty, 0 \le R \le 1, -\infty < \xi_0 < +\infty.$

Point mapping of hyper plane Π into itself is written in the form (2.1.16). By the values k_1, k_2, η_i, it should be meant that:

$$k_1 = \left(\lambda_2^2 - \eta_1^2\right)/\phi\lambda_2^2, k_2 = \left(\lambda_2^2 - \eta_2^2\right)/\phi\lambda_2^2.$$

and η_1, η_2 are defined as the roots of frequency equations.

$$\eta^4 - \left[(1+\phi)\lambda_2^2 + \lambda_1^2\right]\eta^2 + \lambda_1^2\lambda_2^2 = 0. \quad (2.3.3)$$

value Δ is written in the form

$$\Delta = 1 + \lambda_1^2\left(\lambda_2^2 - 1\right) - \lambda_2^2(1+\phi). \quad (2.3.4)$$

To study the stability of single-impact n-fold periodic modes, a characteristic equation is derived, which has the form (2.2.12). Expanding the determinant (2.2.12), we obtain a characteristic polynomial of the fourth degree.

$$\sum_{i=0}^{4} a_i z^{4-i} = 0. \quad (2.3.4)$$

The coefficients of which after a series of simplifications will have the form,

$$\begin{aligned}
a_0 &= \eta_1\eta_2\frac{(k_2-k_1)^2}{k_1k_2}\dot{\xi}_0, a_4 = R^2 a_0,\\
a_1 &= \left\{\sum_{i=1}^{6} Q_i\right\}\dot{\xi}_0 + \left\{\sum_{i=1}^{2} P_i\right\}\xi_0,\\
a_2 &= \left\{\sum_{i=1}^{3} d_i + \sum_{i=1}^{4} Q_{i+6}\right\}\dot{\xi}_0 + \left\{\sum_{i=1}^{4} P_{i+2}\right\}\xi_0,\\
a_3 &= \left\{\sum_{i=1}^{6} Q_{i+10}\right\}\dot{\xi}_0 + \left\{\sum_{i=1}^{2} P_{i+6}\right\}\xi_0.
\end{aligned} \quad (2.3.5)$$

Included in them d_i, P_i, Q_i are expressed in terms of the system parameters as follows:

$$\begin{aligned}
d_1 &= -2\eta_1\eta_2\left[(1-R)^2 + 2R(k_1/k_2 - k_2/k_1)\right]\cos T\eta_1 \cdot \cos T\eta_2,\\
d_2 &= -\eta_1\eta_2(1+R)^2(\eta_1/\eta_2 + \eta_2/\eta_1)\sin T\eta_1 \cdot \sin T\eta_2,\\
d_3 &= \eta_1\eta_2\left[4R + (1+R)^2(k_1/k_2 + k_2/k_1)\right],\\
Q_1 &= 2f(R/k_1 + 1/k_2)\cos T\eta_2,
\end{aligned}$$

$$
\begin{aligned}
Q_3 &= \frac{1+R}{2a}\left[(a-bk_1)\left(\lambda_2^2-\lambda_2^2R/k_1-1\right)+k_1\right]\sin T\eta_2 ctg\frac{T}{2}\eta_1,\\
Q_5 &= \frac{1+R}{a}\frac{\eta_2}{\eta_1}\left[\frac{(k_2-k_1)a\eta_1^2}{k_2}+\frac{k_1(a-bk_2)}{k_2}\right.\\
&\quad\left.-\lambda_2^2(a-bk_1)\left(1-\frac{R}{k_2}\right)-k_1\right]\cos^2\frac{T}{2}\eta_1,\\
Q_7 &= \frac{1+R}{2a}\left[2a-b(1-R)k_1+\lambda_2^2(a-bk_1)\left(R-1+\frac{2R}{k_1}\right)\right.\\
&\quad\left.-(1-R)k_1-Ra\eta_1^2-a\eta_1^2\right]\cos T\eta_1\sin T\eta_2 ctg\frac{T}{2}\eta_1,\\
Q_9 &= \frac{1+R}{a}\frac{\eta_1}{\eta_2}\left[\frac{k_2}{k_1}(-2a-bk_1R-bk_1)-\lambda_2^2(a-bk_2)\left(R-1+\frac{2R}{k_1}\right)\right.\\
&\quad\left.+a\eta_2^2(R-1)+k_2(1-R)+\frac{2ak_2\eta_2^2}{k_1}\right]\cos T\eta_1\cos^2\frac{T}{2}\eta_2,\\
Q_{11} &= \frac{1+R}{2}\left[(1+R)\eta_1^2-1-\frac{Rk_1}{a}(1+b)-\left(1+\frac{1}{k_1}\right)\frac{R\lambda_2^2(a-bk_1)}{a}\right]\\
&\quad\sin T\eta_2 ctg\frac{T}{2}\eta_1,\\
Q_{13} &= (1+R)\frac{\eta_1}{\eta_2}\left[\frac{R\lambda_2^2}{a}(a-bk_2)\left(1+\frac{1}{k_1}\right)-\eta_2^2\left(R+\frac{k_2}{k_1}\right)\right.\\
&\quad\left.+R\frac{k_2}{a}(1+b)+\frac{k_2}{k_1}\right]\cos\frac{T}{2}\eta_2,\\
Q_{15} &= 2Rf\left[\frac{R}{k_2}+\frac{1}{k_1}\right]\cos T\eta_2.
\end{aligned}
\tag{2.3.6}
$$

$$
\begin{aligned}
P_1 &= \frac{k_2-k_1}{a}\eta_1\left[\lambda_2^2\left(1-\frac{R}{k_1}\right)(b-\phi a)-1+\frac{a}{k_1}(1+R\lambda_1^2)-b\right]\sin T\eta_2,\\
P_3 &= 2\frac{k_2-k_1}{a}\left[a+R(-\lambda_2^2b+\lambda_2^2\phi a+\lambda_1^2a)\right]\frac{\eta_2}{k_2}\sin T\eta_1\cos T\eta_2,\\
P_5 &= \frac{k_2-k_1}{a}\left[b-\lambda_2^2b+\lambda_2^2\phi a+1\right](1-R)\eta_1\sin T\eta_2\cos T\eta_1,\\
P_7 &= \frac{k_2-k_1}{a}\eta_1[1+B]R+\frac{R\lambda_1^2a}{k_1}+\frac{a}{k_1}\\
&\quad-R\lambda_2^2(b-\phi a)\left(1+\frac{1}{k_1}\right)\sin T\eta_2.
\end{aligned}
$$

The expressions for P_i, Q_i with even indices are obtained from the relations, for P_i, Q_i with odd indices by point transformation

$$
\begin{aligned}
Q_{2j} &= \Pi_{1,2}Q_{2j}-1,\ (j=1,2,\ldots,8),\\
P_{2i} &= \Pi_{1,2}P_{2i}-1,\ (i=1,2,3,4).
\end{aligned}
\tag{2.3.7}
$$

where $\Pi_{1,2}$—transposition operator that changes indices k and η in P and Q from 1 to 2 or from 2 to1depending on the indices with which they appear in these expressions.

To construct the regions of N-stability, we find expressions for the boundaries N_+, N_-, N_φ in terms of the characteristic polynomial. Assuming in (2.3.4) $z = +1, z = -1$, we obtain the stability boundaries N_+, N_- in the parameter space in the following form

$$\sum_{i=0}^{4} a_i = 0, \sum_{i=0}^{4} (-1)^i a_i = 0. \tag{2.3.8}$$

Assuming further $z = e^{\pm i\varphi}$, we find the boundary N_φ that splits into 4 surfaces[2]

$$\begin{aligned} 2(a_0 - a_4) + (a_1 - a_3) &= 0, \\ 2(a_0 - a_4) - (a_1 - a_3) &= 0, \\ 3(a_0 + a_4) - a_2 &= 0, \\ (a_0 - a_4)^2(a_0 + a_4 - a_2) - (a_1 - a_3)(a_1 a_4 - a_0 a_3) &= 0. \end{aligned} \tag{2.3.9}$$

The part of boundaries which determine the stability region of a single-impact periodic motion (N_+) coincide with the boundary of existence of periodic motion. When passing through the boundary surface N_-, the periodic motion does not disappear, but loses its stability. However, along with this, a single-impact motion turns into two, three, etc., impact periodic motion when passing through N_-.

The construction of stability regions along the boundaries (2.3.8) and (2.3.9) is rather difficult, as it has been found further that the stability region is not a singly connected domain and this significantly complicates the construction of boundaries (2.3.8), (2.3.9) by solving transcendental equations. However, we can avoid this difficulty in the following way: as it is known [11–13] that for stability of periodic mode, it is necessary and sufficient that all the roots of the characteristic Eq. (2.3.4) lie inside a circle of a unit radius, i.e., the condition is held

$$|z_i| < 1, (i = 1, 2, 3, 4). \tag{2.3.10}$$

Verification of conditions (2.3.10) is also laborious, since it requires solving a 4th degree equation. It is known that condition (2.3.10) can be verified by simple Hurwitz inequalities. For this, we make a conformal mapping of a circle of unit radius onto the left half-plane using the transformation

$$W = \frac{z+1}{z-1}. \tag{2.3.11}$$

[2] Equations (2.3.8) and (2.3.9) coincide with the corresponding expressions in [35]. Note that only the last Eq. (2.3.9) can be employed to construct the boundary using the shading rule [36, 37].

After that, the coefficients $a_i (i = 0, 1, \ldots, 4)$ will take the following form (we denote them by B_j) [38]:

$$B_j = \sum_{k=0}^{m} a_k \left[\binom{m-k}{j} - \binom{k}{1}\binom{m-k}{j-1} + \binom{k}{2}\binom{m-k}{j-2} - \ldots \right.$$
$$\left. +(-1)^{j-1}\binom{k}{j-1}\binom{m-k}{1} + (-1)^j \binom{k}{j} \right]. \tag{2.3.12}$$

where

- m is degree of characteristic polynomial,
- j is a sequence number of coefficients of the transformed characteristic polynomial,
- k is a sequence number of coefficients of the specified characteristic polynomial.

After recalculating the coefficients a_i with the help of formulas (2.3.12), the stability problem is solved using the Hurwitz criterion, which significantly saves the time for parameter space partitioning into stability regions.

Using the given above formulas, the stability boundaries of a single-impact periodic motion were constructed. Figs. 2.3 and 2.4 shows in the plane λ_1, ξ_0 the stability regions (N-stability), hatching from the left up to the right of single-impact motion for $n = 1.2$ at equal masses m_1, m_2 and equal stiffnesses of the lower and upper springs $C_1 = C_2$. Dotted line corresponds to the boundary of existence, found using the formulas (2.2.7), (2.2.8). Chain-dotted line indicates the boundary of the region of impact less motions $\xi_0 > |a|$.

It can be seen from Figs. 2.3, 2.4 and 2.5 that a single-impact motion is unstable in a significant part of the existence region, thus confirming the importance of the research into stability of periodic motion. The regions, located near the resonant frequencies of the corresponding linear system without a limiter ($\lambda = 0.618$ and $\lambda = 1.618$), are appeared to be the most essential. From these figures, it follows that with increasing multiplicity of period T to the period of external force, the stability regions become much narrower even for λ_1.

In N-stability regions, the magnitude of impact velocity was studied. Thus, in all figures later on, the numbers indicate the magnitude of impact velocity $\dot{\xi}_0$. The figures show that the magnitude of impact velocity in the near resonant frequency regions significantly exceeds that of the regions located far from resonant frequencies.

In Figs. 2.5 and 2.6, stability regions of a single-impact motion are constructed on the plane λ_1, ξ_0 for the case when the vibrator mass is twice as large and twice as less as that of the hammer, respectively. The stiffnesses of the upper and lower springs remain equal ($C_1 = C_2$). Replacing in (2.3.3) λ_2^2 with λ_1^2/ϕ, the equation for determining the resonance frequencies of the corresponding linear model will be rewritten as follows

$$\lambda_1^4 - (1 + 2\phi)\lambda_1^2 + \phi = 0. \tag{2.3.13}$$

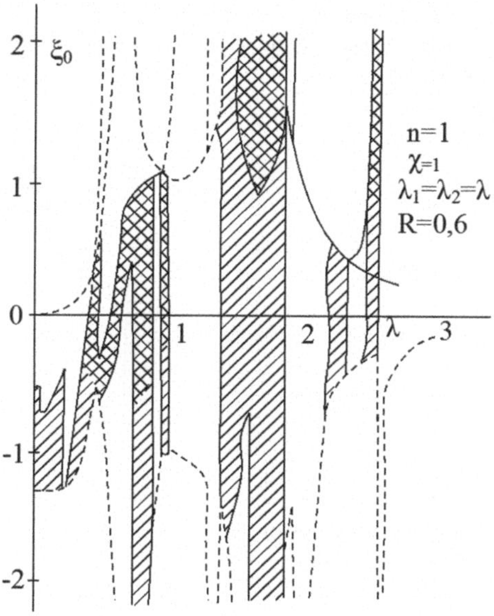

Fig. 2.3 Regions of existence and stability of periodic mode of motion for $n = 1, \chi = 1, \lambda_1 = \lambda_2 = \lambda, R = 0.6$

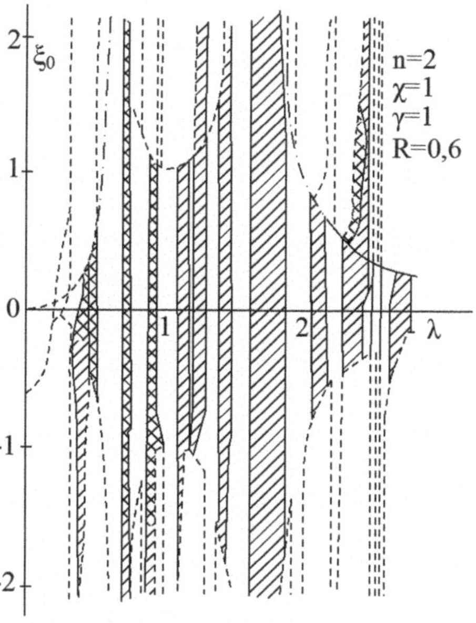

Fig. 2.4 Regions of existence and stability of periodic mode of motion for $n = 2, \chi = 1, \lambda_1 = \lambda_2 = \lambda, R = 0.6$

From the frequency Eq. (2.3.13), we find the expression for the derivatives $d\lambda_{1,2}^2/d\phi$, where λ_1^2 corresponds to larger root and λ_2^2 to smaller root of Eq. (2.3.13) which have the form

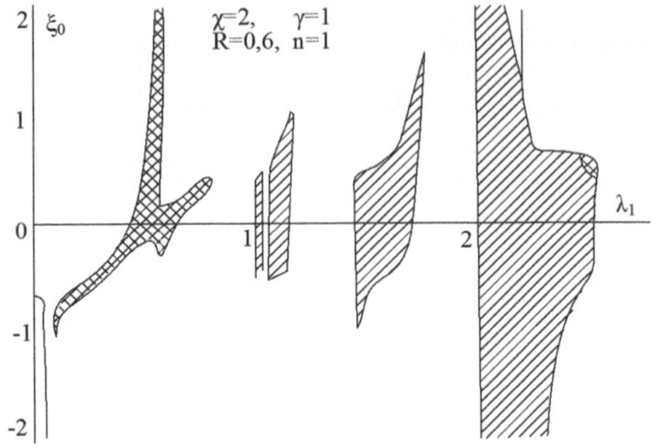

Fig. 2.5 Regions of existence and stability of periodic mode of motion for $n = 1, \chi = 2, \gamma = 1, R = 0.6$

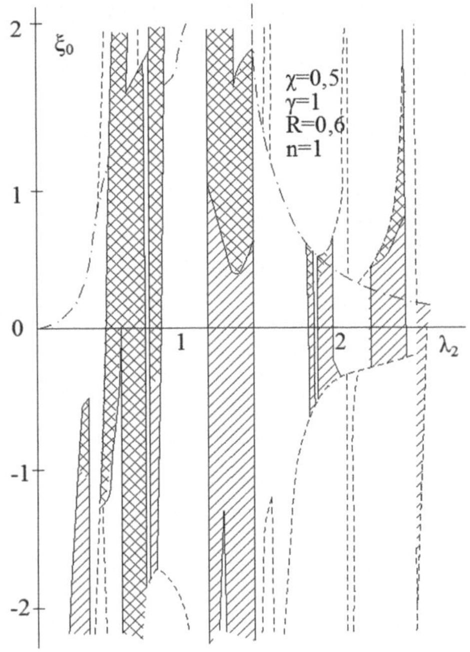

Fig. 2.6 Regions of existence and stability of periodic mode of motion for $n = 1, \chi = 0.5, \gamma = 1, R = 0.6$

$$\frac{d\lambda_{1,2}^2}{d\phi} = 1 \pm \frac{2\phi}{\sqrt{1 + 4\phi^2}}.$$

From these equations, we find that both derivatives are positive. Therefore, we can say that the values of the resonant frequencies with increasing ϕ shift to the region of large λ_1 and since the widest stability regions lie within these frequencies, then they shift to the region of larger λ_1 values with increasing vibrator mass (or decreasing striker mass).

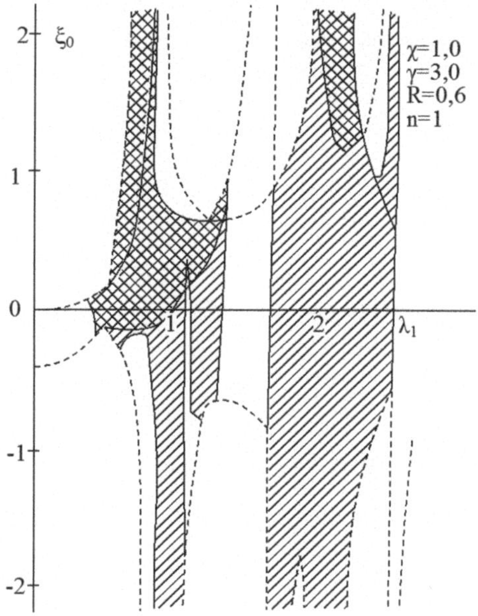

Fig. 2.7 Regions of existence and stability of periodic mode of motion for $n = 1, \chi = 1, \gamma = 3, R = 0.6$

In addition, we note that with growth ϕ the stability regions expand over λ_1 at $\lambda_1 > 1$ (after resonance region) and decrease at $\lambda_1 < 1$ (below resonance region). In this case, the values of impact velocities are seen to increase (see Figs. 2.3 and 2.6).

In Figs. 2.7 and 2.8, similar plottings are performed for the case of equal masses, but different stiffnesses of the upper and lower springs. The ratio $\zeta = C_1/C_2$ equals 0.5 and 3.0, respectively. In this case, the equation for resonant frequencies has the form

$$\lambda_1^4 - (2 + \zeta)\lambda_1^2 + \zeta = 0. \qquad (2.3.14)$$

We determine the derivatives from this equation

$$\frac{d\lambda_{1,2}^2}{d\zeta} = \frac{1}{2}\left[1 \pm \frac{\zeta}{\sqrt{\zeta^2 + 4}}\right].$$

And in this case, for any parameter ζ, the derivatives $d\lambda_1^2/d\zeta$ and $d\lambda_2^2/d\zeta$ are positive. This means that with increasing ζ (with decreasing stiffness of the upper springs or with increasing stiffness of the lower springs) the values of the resonant frequencies shift to the region of large values λ_1. Thus, at $\zeta = 0.5$, $\lambda_2 = 0.468$, $\lambda_1 = 1.51$ and at $\zeta = 3.0$, $\lambda_2 = 1.13$, $\lambda_1 = 2.03$.

The main parts of the stability region of unidirectional periodic motion are located in the vicinity of these resonant frequencies. Significant expansions of stability regions are observed over λ_1 with increasing parameter value ζ.

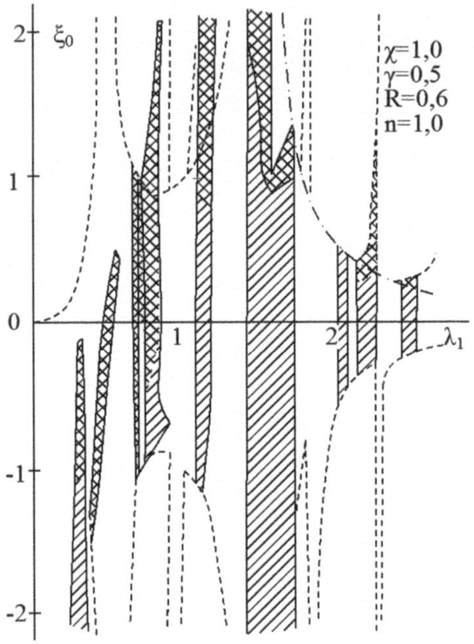

Fig. 2.8 Regions of existence and stability of periodic mode of motion for $n = 1, \chi = 1, \gamma = 0.5, R = 0.6$

Figures 2.9, and 2.10 present the stability regions plotted in the plane λ, ξ_0 for like masses and like stiffnesses for the values of the coefficient of restitution equal to 0.01 and 0.9, respectively. It can be seen from the figures that with increasing the coefficient of restitution, the stability regions become more indented and the cuts in stability region appear from the side of negative gaps.

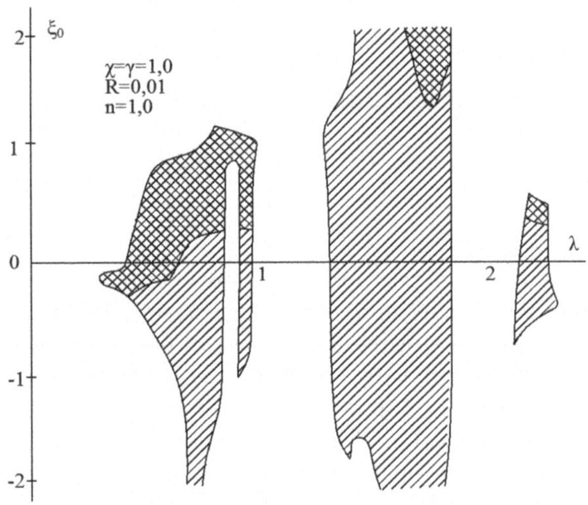

Fig. 2.9 Regions of existence and stability of periodic mode of motion for $n = 1, \chi = 1, \gamma = 1, R = 0.01$

Fig. 2.10 Regions of existence and stability of periodic mode of motion for $n = 1, \chi = 1, \gamma = 1, R = 0.9$

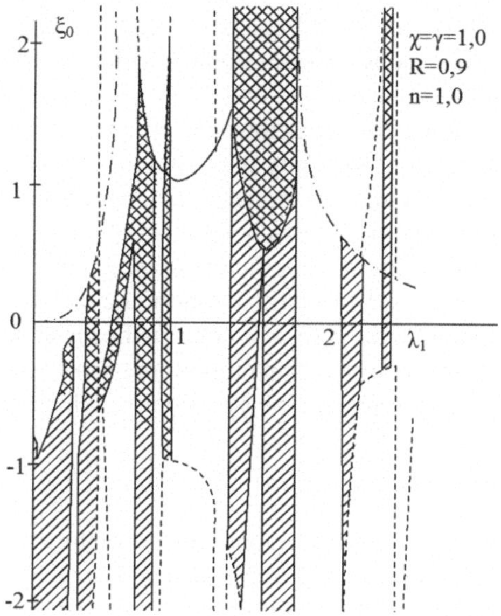

From the results obtained above, it is clear that among all stability regions (in the parameter plane λ, ξ_0), the two widest ones over λ on stand out. These regions lie near the resonance frequencies of the corresponding two-mass linear model. In addition, they are also characterized by the largest values of impact velocities. Therefore, if you tune the system to the optimal mode, taking the maximum pulse as an optimality criterion, then it is necessary to know the location and change of these regions as a function of parameters of the given domains.

However, unidirectional periodic monitoring inspects the entire N-stability region. A single-impact mode is violated when passing through bifurcation curve C_τ, and the equation of which is written in the form (2.2.14)–(2.2.18).

To construct boundary C_τ we proceed as follows: we derive β from Eq. (2.2.15) for fixed values R, ϕ, ζ by setting $\lambda_1^i = \lambda_0 + ih$ where h is λ_1 step and the corresponding gap value ξ_0 is derived from the Eq. (2.2.7). It should be noted that the Eq. (2.2.15) can have several roots $\beta_k \in [0, 2\pi](k = 1, 2, \ldots, \alpha)$. From all α roots of the Eq. (2.2.15), it is necessary to take that value β^* which yields the largest value ξ_0. Extra roots appear due to the fact that in Eqs. (2.2.15) and (2.2.17) the condition $\xi_1 < \xi_0$ is not accounted for.

After plotting the boundary C_τ, it turned out (see Figs. 2.3, 2.4, 2.5, 2.6, 2.7, 2.8, 2.9, and 2.10) that the region of stability of a single-impact motion bounded by N_+, N_-, N_φ surfaces and the bifurcation curve C_τ (part of the stability region with double hatching) is much smaller than the region bounded only by N-surfaces. Thus, to tune the system for a single-impact periodic mode the knowledge of the location of N-stability regions is not enough and it is necessary to indicate that part of the N-stability region which is cut off by the bifurcation surface C_τ.

It is seen from Figs. 2.3, 2.4, 2.5, 2.6, 2.7, 2.8, 2.9, and 2.10 that the location of these regions changes slightly with the gap change, and for small gaps, for example $|\xi_0| < 1$, the stability regions practically remain unchanged. Therefore, further investigation of these regions can be carried out at a constant gap ξ_0. For simplicity of calculation, we study the behavior of these stability regions at zero gap depending on the parameters ϕ and ζ, i.e., depending on the magnitude of the mass ratio and the ratio of stiffness coefficients of the upper and lower springs for different values of the restitution coefficient upon impact equal to $R = 0.2$ and $R = 0.6$.

Figures 2.11 and 2.12 show the stability regions (the shaded part of the plane) plotted for the case $\phi = \text{const}$ in the plane of two parameters λ_1, ζ at $R = 0.2$ and $R = 0.6$, respectively. The dashed line shows the resonance curves of the corresponding linear system, and the equations of which are of the form (see formula (2.3.14))

$$\lambda_1^4 - (2 + \zeta)\lambda_1^2 + \zeta = 0.$$

The stability regions corresponding to a higher resonance frequency, when changing ζ at $R = 0.2$ and $R = 0.6$, are located around this resonance frequency, and for small ζ stability regions are strongly indented. The low-frequency stability region at $R = 0.2$ lies for $\zeta > 0.8$ in the vicinity of this resonance frequency and for $\zeta < 0.8$ partition of the region occurs. Moreover, in the vicinity of this resonance frequency for small ζ, stability regions are not observed as they shift to the region of large ζ. Thus, for selecting the parameters of this stability region, it is necessary to reduce the frequency of external force with decreasing ζ.

If the coefficient of restitution equals 0.6, the lower frequency stability regions slightly shift to the regions of lower λ with increasing ζ. For small ζ at $\xi_0 = 0$, the

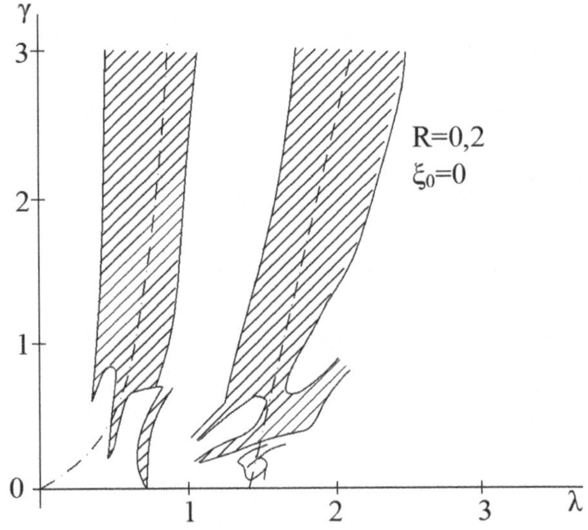

Fig. 2.11 Regions of existence and stability of periodic mode of motion for $n = 1, \chi = 1, R = 0.2, \xi_0 = 0$

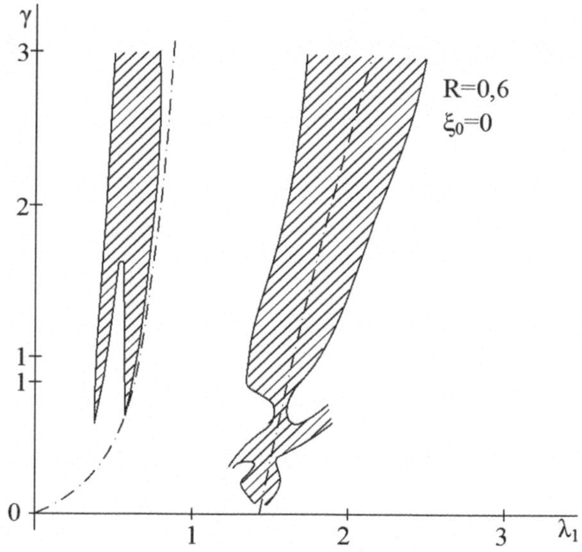

Fig. 2.12 Regions of existence and stability of periodic mode of motion $n = 1, \chi = 1, R = 0.6, \xi_0 = 0$

stability region in the vicinity of the resonant frequency is not observed, in this case a negative gap is to be chosen.

Figures 2.13 and 2.14 present the stability regions plotted in the parameter plane λ, ϕ^{-1} for the coefficient of restitution equal to 0.2 and 0.6. The dashed line shows the resonance curves of the liver system, and the equations of which in this case are of the form (2.3.13).

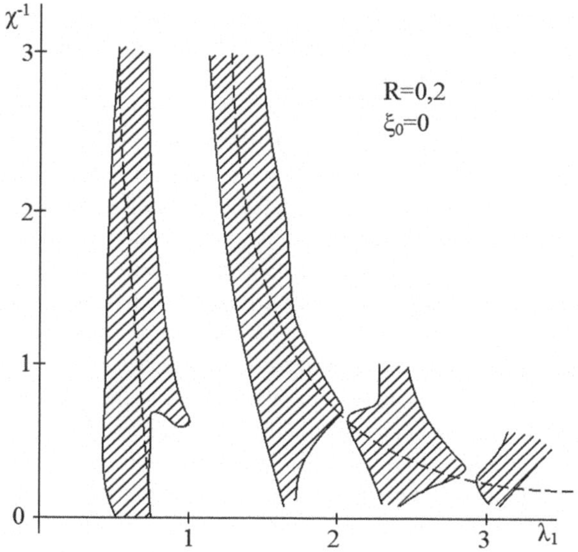

Fig. 2.13 Regions of existence and stability of periodic mode of motion for $n = 1, \gamma = 1, R = 0.2, \xi_0 = 0$

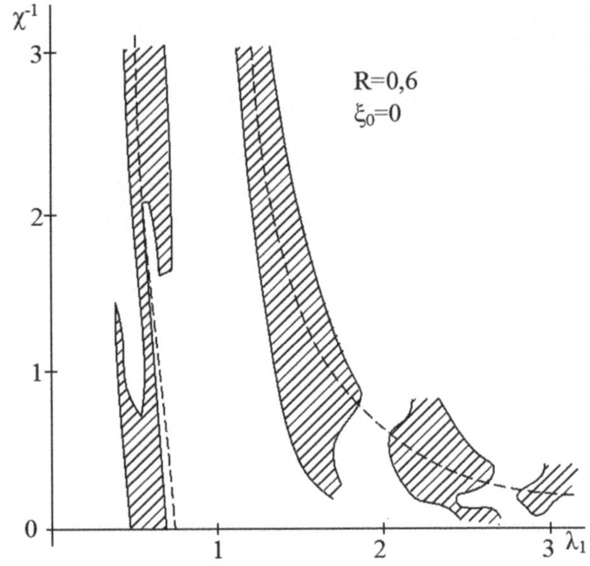

Fig. 2.14 Regions of existence and stability of periodic mode of motion for $n = 1, \gamma = 1, R = 0.6, \xi_0 = 0$

The stability regions at $R = 0.2$ and $R = 0.6$ lie in the vicinity of higher resonant frequency. For small ϕ^{-1} (with a smaller striker mass), the stability region splits into a number of isolated regions. These stability regions seem to continue each other: where the one region ends almost immediately the other begins, etc. The stability regions lying in the vicinity of the resonance frequency are the broadest, and they are getting narrower with distance from the resonance curve.

The lower frequency stability regions at $R = 0.2$ and $R = 0.6$ shift to the right over λ with decreasing ϕ (the mass of the striker is greater than that of the vibrator), i.e., in order to select the parameters belonging to this region, it is necessary to reduce the external force frequency ω.

It can be seen from Figs. 2.9, 2.10, 2.11, 2.12, 2.13 and 2.14 that with increase in the coefficient of restitution R, the stability regions become more indented especially from the side of negative gaps (see Figs. 2.9, 2.10, 2.11, 2.12, 2.13, and 2.14). As a result, the similar location dependences of stability regions on system parameters ϕ and ζ for large R values can be plotted only for large gap values ξ_0.

From the results obtained above, it is seen that the largest stability region lies in the vicinity of higher resonance frequency whatever changes in the system parameters. Studies of the magnitude of the impact velocity have also shown that the stability region lying in the vicinity of a higher resonance frequency has also the greatest value $\dot{\xi}_0$. Therefore, to tune the system to the optimal mode, if we take not only the maximum impact pulse and the magnitude of the system stability margin as the criterion of optimality, we must apparently choose the system parameters in the vicinity of a higher resonant frequency, for which it is necessary to have a low-frequency vibrator. This result does not contradict to operating vibro-impact systems (e.g., a self-propelled vibro-impact machine designed for compaction of a crushed stone prism of a railway bed, the vibrations of which are excited by N. A. Karpov's low-frequency combined self-synchronizing unbalanced vibrator designed at the Central Research Institute of Railways [39]).

The study of more complex modes and some theoretical verification of the results obtained in this section can be carried out using electronic modeling. A schematic diagram for modeling vibro-impact systems was proposed in a number of works: Bebikhov [35] (for single-mass systems), Peterka [20, 21], Goldin [23, 24], Althausen, Godes [37, 40] and in many other works (for two-mass systems), and multi-mass [41].

The general block diagram simulating a two-mass vibration striker described by Eqs. (2.3.1), (2.3.2) is shown in Fig. 2.15. For the convenience of modeling, the initial system of differential equations

$$\xi_{\text{HO6}} = 40\xi_{\text{cm}},\ \tau_{\text{HO6}} = 2\tau_{\text{cm}}.$$

was reduced by introducing new variables to a system of equations of the form at $\xi_1 < \xi_0$

$$\ddot{\xi}_1 = -\frac{\lambda_2^2}{4}(\phi\xi_1 - \xi_2) - \frac{\lambda_1^2}{4}\xi_1.$$

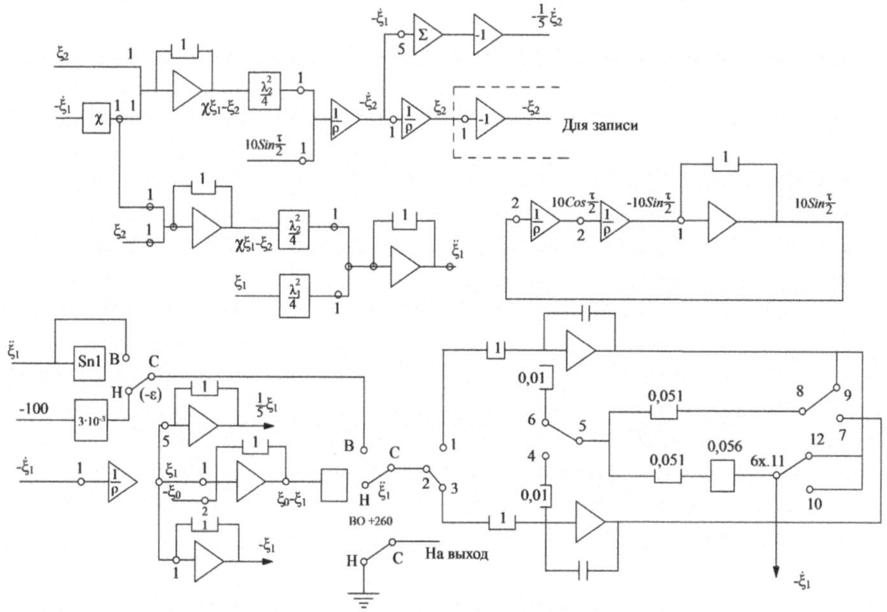

Fig. 2.15 Schematic diagram for solving a system of differential Eqs. (2.3.1) and (2.3.2) on an analog computer MH-8

$$\ddot{\xi}_2 = -\frac{\lambda_2^2}{4}(\xi_2 - \phi\xi_1) + 10\sin\frac{\tau}{2}. \quad (2.3.15)$$

at $\xi_1 = \xi_0$

$$\dot{\xi}_1 = -R\dot{\xi}_0.$$

These equations were simulated for the coefficient of restitution equal to 0.6. The first task was to compare the theoretical boundaries C_τ obtained from formulas (2.2.14)–(2.2.18) with the boundaries obtained by modeling on a continuous device. To improve convergence to periodic motion, the initial conditions on integrators were chosen according to formulas (2.2.3)–(2.2.6).

From the results obtained during the simulation, it turned out that the boundary C_τ coincides fairly well with the theoretical boundary (2.2.3)–(2.2.6). Figure 2.16a–c presents the motion oscillograms for the striker and vibrator as well as a change in their velocities for the parameter values $\lambda_1 = \lambda_2 = 1.6$, $\phi = 1$, ξ_0, respectively, equal to 0.4, 0.9, and 2.0. The gap value of 0.9 is bifurcational since with its decrease, a single-impact periodic mode switches into two, three, etc., impact mode. It is clearly seen from Fig. 2.16b that at the moment $\tau = \tau*$ the value of impact velocity equals zero. For the gap value equal to 0.4, a three-impact periodic mode was established. It should be noted that with a decrease in the gap at a constant frequency of the driving force ω, the magnitudes of the oscillation amplitudes of the impact mass and impact velocity decrease.

Further we attempt to divide stability region N into the modes of various qualitative behavior. We present the simulation results for the widest stability region lying in the vicinity of a higher resonance frequency for $R = 0.6 : \phi = 1.0$, $\zeta = 1.0$. Figure 2.17 shows the partition of the plane parameter λ_1, ξ_0 into regions of various qualitative behavior for positive gaps.

For parameter values higher than parameters determined from the boundary C_τ, there is a single-impact stable single mode with a period $T = 2\pi$. With gap decreasing, a single-impact periodic regime transforms into a double-impact (region D_{21}), then into a three-impact (region D_{31}), etc., i.e., an increase in the number m with gap ξ_0 decreasing is observed. For negative gaps in the vicinity of a higher resonant frequency, there exist periodic motions with m greater than 4. Apparently, domains D_{i1} can include D_{ij} domains, where j > 1, but such modes were not distinguished in the domains. Modes with stops were not considered as well.

Figures 2.18 and 2.19 present the motion oscillograms of impact mass in the selected domains D_{i1} for various frequencies of the driving force ω and the gap ξ_0.

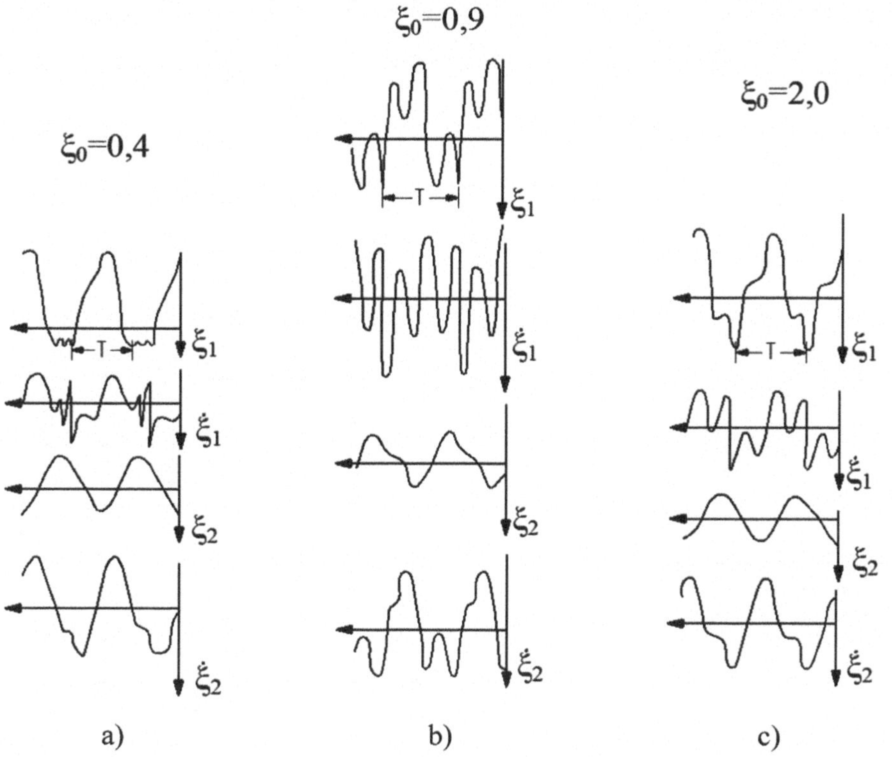

Fig. 2.16 Oscillograms for various values of initial gap of mechanism, calculated on an MH-8 analog computer

Fig. 2.17 Stability areas with one shock per period (D_{11}), two shocks per period (D_{21}), three (D_{31}), four (D_{41})

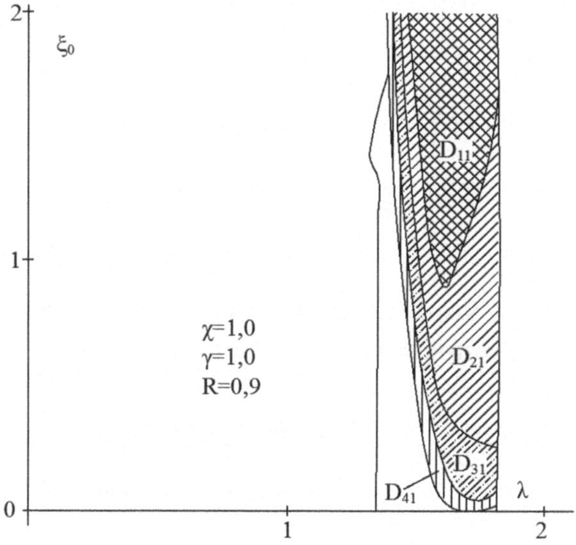

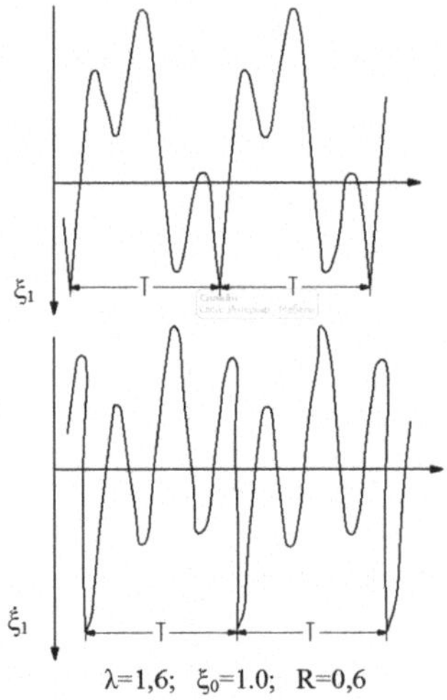

Fig. 2.18 Oscillograms for various gap values ξ_0 and frequencies ω in stability regions D_{i1} for $n = 1, \chi = 1{,}6, \gamma = 1, R = 0.6, \xi_0 = 1$

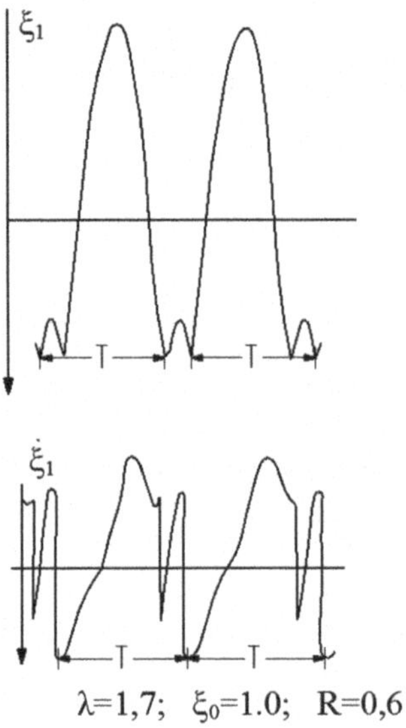

Fig. 2.19 Oscillograms for various gap values ξ_0 and frequencies ω in stability regions D_{i1} for $n = 1, \chi = 1$, $\gamma = 1.7, R = 0.6, \xi_0 = 0$

2.4 A Study of the Stability of a Two-Mass Non-spring-loaded Vibration Hammer

In this section, we study a two-mass vibro-impact system used for compaction of soil, gravel, and concrete [42], for a pile vibro-impact immersion [23, 24], as well as in other mechanisms [1, 2]. The design of such and similar systems is facilitated by the fact that they are not connected with the processed material through elastic (rigid) bonds, and the mechanism interacts with the processed substance through impacts.

When using such a system as a vibro-impact compaction, the coefficient of restitution upon impact R can vary from about 0 to 0.4, and when using it as a pile vibro-impact driver, the coefficient R varies from 0.2 to 0.6 [22, 43].

The considered model of vibro-impact compactor consists of the mass m_2 (vibrator), elastically fixed with the elasticity coefficient C_3 and affected by an external periodic force $F \sin \omega t$, and the mass m_1 (striker) elastically connected with a vibrator by the stiffness coefficient C_2.

In dimensionless variables

$$\xi_1 = \frac{m_2\omega^2}{F}\left(x_1 - \frac{P_1}{c_2} - \frac{P_1 + P_2}{c_3}\right),$$
$$\xi_2 = \frac{m_2\omega^2}{F}\left(x_1 - \frac{P_1 + P_2}{c_3}\right). \tag{2.4.1}$$

The equations of motion of the striker and vibrator in the intervals between impacts $\xi_1 < \xi_0$ are written in the form

$$\ddot{\xi}_1 + \lambda_1^2(\xi_1 - x\xi_2) = 0,$$
$$\ddot{\xi}_2 + \lambda_1^2(x\xi_2 - \xi_1) + \lambda_2^2\xi_2 = \sin \tau. \tag{2.4.2}$$

where the following notations are introduced

$$\lambda_1^2 = \frac{C_2}{m_1\omega^2}, \lambda_2^2 = \frac{C_3}{m_2\omega^2}, x = \frac{m_1}{m_2}. \tag{2.4.3}$$

Note that if in general model, the parameter x was denoted by the ratio of the upper mass to the lower mass $x = \frac{m_2}{m_1}$, then in this model in order to preserve the general form of differential Eq. (2.4.2), the dimensionless parameter x is reciprocal, equal to the ratio of the lower mass to the upper one.

The phase space of the system under consideration is five-dimensional. As in general case, it is divided into two regions $\xi_1 < \xi_0$ and $\xi_1 \geq \xi_0$ by a hyperplane Π $(\xi_1 = \xi_0)$. All motions of the system lie in the region of phase space determined by inequality $\xi_1 \leq \xi_0$. Therefore, during the impact motions of the system, investigation of these motions is reduced to studying the point mapping of the hyperplane Π into itself. The point mapping of the hyperplane into itself has the form (2.2.2), in which the distribution coefficients K_i included in expressions (2.2.2) are determined from the relations

$$K_i = \left(\lambda_2^2 + \lambda_1^2 x - \eta_i^2\right)/\lambda_1^2, (i = 1, 2). \tag{2.4.4}$$

where η_1 and η_2 are the roots of frequency equations

$$\eta_4 - \left[\lambda_2^2 + \lambda_1^2(1 + x)\right]\eta^2 + \lambda_1^2\lambda_2^2 = 0. \tag{2.4.5}$$

The amplitude of the forced oscillations of the upper and lower masses is written as

$$\eta_4 - \left[\lambda_2^2 + \lambda_1^2(1+x)\right]\eta^2 + \lambda_1^2\lambda_2^2 = 0. \tag{2.4.6}$$

where $\Delta = \left(1 - \eta_1^2\right)\left(1 - \eta_2^2\right)$.

We will find the coordinates of the fixed point corresponding to single-impact periodic motion of the point mapping S from the system (2.2.2), assuming $\dot{\xi}_2 = \dot{\xi}_2^0$, $\tau_1 - \tau_0 = T$. Solving the system for any integer n with respect to $\dot{\xi}_0$, ξ_2^0, $\dot{\xi}_2^0$, $\sin\tau_0$, $\cos\tau_0$, we obtain

$$\xi_1 = \frac{m_2\omega^2}{F}\left(x_1 - \frac{P_1}{c_2} - \frac{P_1 + P_2}{c_3}\right),$$
$$\xi_2 = \frac{m_2\omega^2}{F}\left(x_1 - \frac{P_1 + P_2}{c_3}\right). \tag{2.4.7}$$

where

$$\eta_4 - \left[\lambda_2^2 + \lambda_1^2(1+x)\right]\eta^2 + \lambda_1^2\lambda_2^2 = 0. \tag{2.4.8}$$

and value g_2 is derived from g_1 by replacing the K indices from 1 to 2 and from 2 to 1, as well as by replacing η_1 with η_2. Note that the coordinates of the fixed point (2.4.7), (2.4.8) can be also written in the form (2.2.3)–(2.2.6). However, by the quantities K_1,K_2,a, b, η_1 and η_2, we should understand the above introduced notations (2.4.4)–(2.4.6). The equation for finding the impact velocity is derived from (2.4.7) using the relation $\sin^2\tau_0 + \cos^2\tau_0 = 1$. Inserting into it expressions for $\sin\tau_0$ and $\cos\tau_0$, we obtain

$$\dot{\xi}_0 = \frac{2f}{g}\left\{F\xi_0 + \sqrt{\frac{\lambda_1^4 x^2}{\Delta^2}g - 4\left(1 - R^2\right)f^2\xi_0^2}\right\}. \tag{2.4.9}$$

In (2.4.9), the plus sign is taken before the root. As it will be shown below, if we take the minus sign in front of the root, then one of the necessary conditions for the stability of a fixed point will not be satisfied.

Since the value of the impact velocity must be real and positive, the boundary of existence of a single-impact n-fold periodic motion will be written in the form

$$|\xi_0| = \lambda_1^2 x / |\Delta|,\ F\xi_0 < 0,$$
$$|\xi_0| = \frac{\lambda_1^2 x}{|\Delta|}\sqrt{\frac{F^2}{(1-R)^2 f^2} + 1},\ F\xi_0 > 0. \tag{2.4.10}$$

The boundaries of existence are also line segments $n\eta_1 = K$, $n\eta_2 = \ell$, $n \neq K$, $K \neq \ell$, $K, \ell, n = 1, 2, \ldots$, in which the impact velocity equals to zero.

To study the stability of a fixed point of point mapping S corresponding to a unidirectional periodic motion, we compose a characteristic polynomial. For this, we linearize the point mapping (2.2.2) in a neighborhood of the fixed point (2.4.7), (2.4.9). The characteristic equation $\chi(z) = 0$ can be written as the following determinant

$$\begin{bmatrix} R\eta_2^{-1}\zeta_2 & \kappa_1(z-C_2) & -\kappa_1\eta_2^{-1}\zeta_2 & z(\dot{\xi}_0-\kappa_1\dot{\xi}_2^0)+A_2C_2+\eta_2^{-1}\zeta_2(a-b\kappa_1)\zeta_0-\eta_2B_2\zeta_2-C_2(a-b\kappa_1)C_0 \\ z+RC_1 & \kappa_2\eta_1\zeta_1 & \kappa_2(z-C_1) & z(\ddot{\xi}_1^0-\kappa_2\ddot{\xi}_2^0)-A_1\eta_1\zeta_1+C_1(a-b\kappa_2)\zeta_0-B_1\eta_1^2C_1+\eta_1\zeta_1(a-b\kappa_2)C_0 \\ \zeta_1R\eta_1^{-1} & \kappa_2(z-C_2) & -\kappa_2\eta_1^{-1}\zeta_1 & z(\dot{\xi}_0-\kappa_2\dot{\xi}_2^0)+A_1C_1+\eta_1^{-1}\zeta_1(a-b\kappa_2)\zeta_0-\eta_1B_1\zeta_1-C_1(a-b\kappa_2)C_0 \\ z+RC_2 & \kappa_1\eta_2\zeta_2 & \kappa_1(z-C_2) & z(\ddot{\xi}_1^0-\kappa_1\ddot{\xi}_2^0)-A_2\eta_2\zeta_2+C_2(a-b\kappa_1)\zeta_0-B_2\eta_2^2C_2+\eta_2\zeta_2(a-b\kappa_1)C_0 \end{bmatrix} \tag{2.4.11}$$

The following notations are introduced here

$$C_i = \cos T\eta_i,\ \zeta_i = \sin T\eta_i,$$
$$\zeta_0 = \sin\tau_0,\ C_0 = \cos\tau_0. \tag{2.4.12}$$

Expanding the determinant (2.4.11), we obtain a fourth-degree polynomial with respect to z in the form

$$\sum_{j=0}^{4} B_j z^{4-j} = 0. \tag{2.4.13}$$

where the coefficients of polynomial have the form (2.3.5) and by quantities κ_i, η_i, x, a and b it is necessary to understand the expressions introduced above (2.4.3–2.4.6).

We will show that the motion corresponding to the impact velocity

$$\dot{\xi}_{z0} = \frac{2f}{g}\left\{F\xi_0 - \sqrt{\frac{\lambda_1^4 x^2}{\Delta^2}g - 4\left(1-R^2\right)f^2\xi_0^2}\right\}. \tag{2.4.14}$$

is always unstable. For this, substituting the coefficients of the polynomial (2.4.13) into one of the necessary stability condition $\sum_{i=0}^{4} B_i > 0$ and using relations (2.4.7), (2.4.14) after rather cumbersome transformations, we obtain the following inequality

$$-\sqrt{a^2\left[F^2+(1-R)^2f^2\right]-(1-R)^2f^2\xi_0^2} > 0. \tag{2.4.15}$$

which, obviously, is not satisfied at any values of the system parameters. This means that the motion corresponding to this value of impact velocity is always unstable. Therefore, in (2.4.9), the plus sign is taken before the root. In the phase space, this corresponds to the fact that one of the two fixed points of the point mapping S corresponding to the impact velocity (2.4.14) is always unstable.

For partitioning the parameter space into stability regions of single-impact n-fold periodic modes, we use the technique described in § 2.3, namely: according to formulas (2.3.12), the coefficients of the characteristic polynomial (2.4.13) are transformed to new coefficients after which for the transformed coefficients the Routh–Hurwitz inequalities can be used to determine the stability of a fixed point.

We study the effect of the system parameters x, γ, λ, ξ_0 and R on the location of stability regions. Partitioning will be performed in the plane of the two most interesting parameters ξ_0 and λ for fixed x, γ and R.

Figure 2.21 presents N stability regions (hatching from left up to right) for single-impact periodic motions. Boundary C_τ (dash-dotted line), the equations of which can be written as follows

$$\begin{aligned} &\xi_1\left(\tau^*\right) = \xi_0, \dot{\xi}_1\left(\tau^*\right) = 0, \\ &\tau_0 \le \tau^* \le \tau_1, \tau_1 - \tau_0 = 2\pi n. \end{aligned} \tag{2.4.16}$$

or after some transformations using expressions for the coordinates of a fixed point, Eqs. (2.4.7)–(2.4.9) can be rewritten in the form

$$\Omega_1\Omega_4 - \Omega_2\Omega_3 = 0. \tag{2.4.17}$$

where Ω_i (i =1, 2, 3, 4) are derived from (2.2.12), intercepts a significant part from the stability region. Periodic single-impact motion is possible only in regions with double hatching.

Figures 2.20, 2.21, 2.22, 2.23 and 2.24 present stability regions constructed for x = 0.5; 1.0; 3.0 at fixed values of the remaining parameters $\gamma = 1.0, = 0.2$. Of all the stability regions, as in the case considered in § 2.3, the two widest over λ_1 regions are distinguished in the plane ξ_0, λ_1 lying in the vicinity of the resonant frequencies (R.F.), corresponding to the two-mass linear model (R.F.), the values of which are determined from the equations

$$x\lambda_1^4 - (1 + 2x)\lambda_1^2 + 1 = 0. \tag{2.4.18}$$

Solution to the Eq. (2.4.18) is written in the form

$$\lambda_{1,2}^2 = \left[1 + (2x)^2\right] \pm \sqrt{1 + (2x)^{-1}}. \tag{2.4.19}$$

Hence, it follows that derivatives $\frac{d\lambda_1^2}{dx}$ and $\frac{d\lambda_2^2}{dx}$ in the form

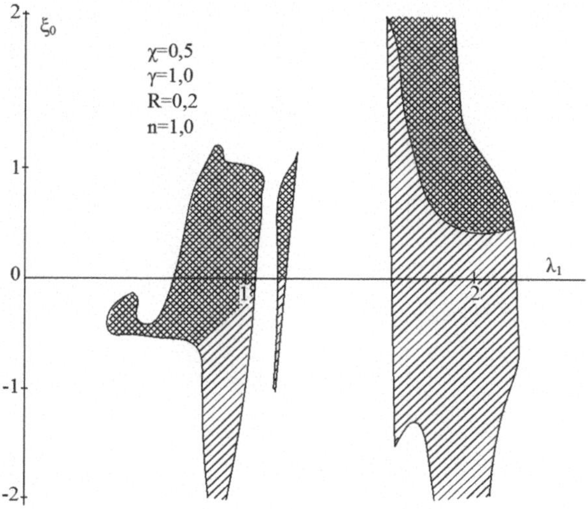

Fig. 2.20 Regions of existence and stability of periodic mode of motion for $\chi = 0.5, \gamma = 1, R = 0.2, n = 1$

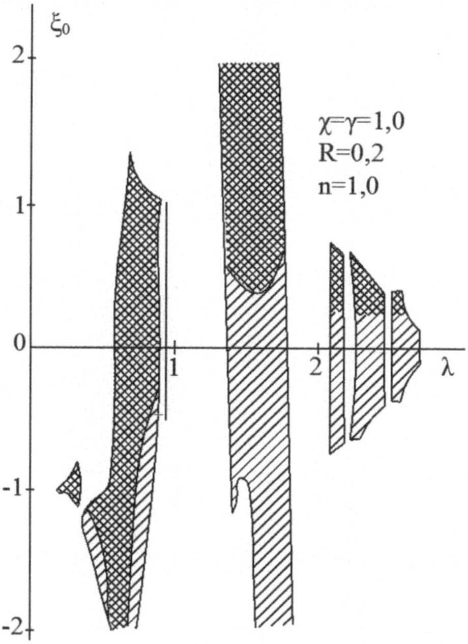

Fig. 2.21 Regions of existence and stability of periodic mode of motion for $\chi = 1, \gamma = 1, R = 0.2, n = 1$

$$\frac{d\lambda_{1,2}^2}{dx} = -(2\text{x})^{-2}\left[2 + \frac{1}{\sqrt{1 + (2\text{x})^{-1}}}\right].$$

for any parameter values x are negative (λ_1^2 and λ_2^2 mean the same as in (2.3.13)). Thus, with increasing x, values (R.F.) decrease, which is equivalent to a shift of the

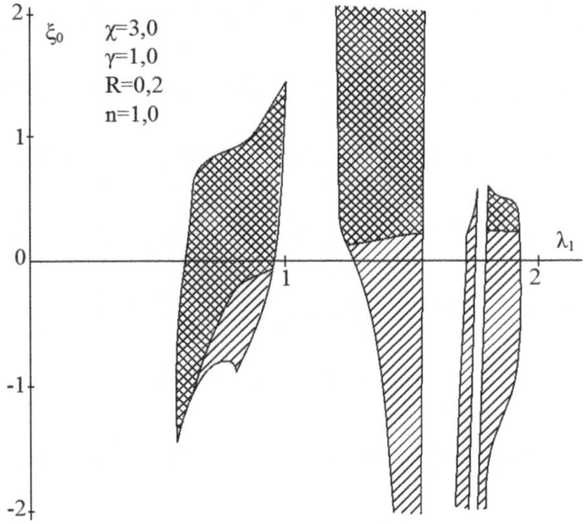

Fig. 2.22 Regions of existence and stability of periodic mode of motion for $\chi = 3, \gamma = 1, R = 0.2, n = 1$

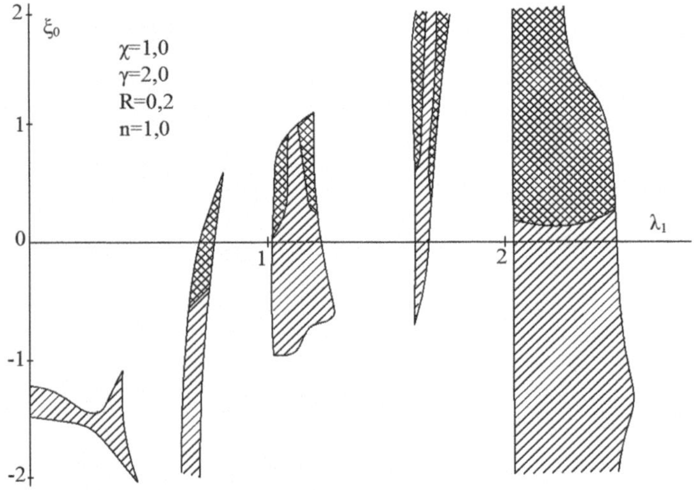

Fig. 2.23 Regions of existence and stability of periodic mode of motion for $\chi = 1, \gamma = 2, R = 0.2, n = 1$

stability regions toward lower λ (or to the region of high frequencies of the driving force ω). Values of (R.F) for $x = 0.5$; 1.0; 3.0 will be 0.765 and 1.848; 0.619 and 1.618; 0.391 and 1.477, respectively.

Figs. 2.23 and 2.24 show the partition of the λ_1, ξ_0 plane into stability regions of single-impact single periodic motions for $\gamma = 2.0$ and 0.5, for the case when

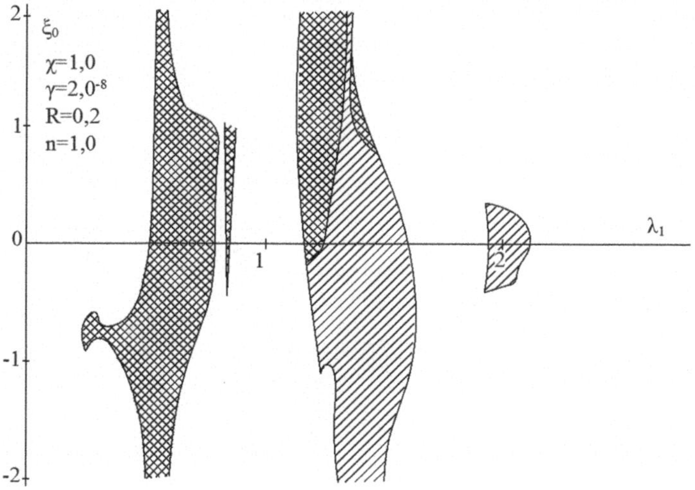

Fig. 2.24 Regions of existence and stability of periodic mode of motion for $\chi = 1$, $\gamma = 0{,}5$, $R = 0.2$, $n = 1$

the mass ratio is 1.0 and the coefficient of restitution R equals 0.2. It is seen from Figs. 2.20, 2.21, 2.22, 2.23 and 2.24 that in the λ_1, ξ_0 plane there are the two widest over λ_1 stability regions in which the values of impact velocities are the highest as compared with velocities in other stability regions. These regions are also located in the neighborhood (R.F), and values of which are derived from the equation

$$\lambda_1^4 - (1 + 2\gamma)\lambda_1^2 + \gamma = 0. \tag{2.4.20}$$

Derivatives $\frac{d\lambda_1^2}{d\gamma}$ and $\frac{d\lambda_1^2}{d\gamma}$, as in (2.3.13), are positive for any parameter γ. Therefore, with an increase in the stiffness ratio of the lower springs to the upper, the ratio of the partial frequency to that of the driving force increases. This suggests that with increasing γ, the stability regions of single-impact periodic regimes shift to the region of large λ (lower values of the driving force frequency). Thus, for γ equal to 0.5 and 2.0, the corresponding values λ_1 will be 0.541; 1.31 and 0.65; 2.103.

Figures 2.20, 2.21, 2.22, 2.23, and 2.24 show that with increasing x, the stability regions expand and shift to the region of lower driving force frequencies, while with increasing γ (decrease in stiffness of the lower springs), the stability regions expand significantly but shift to a region of larger λ.

Figure 2.25 presents the plotted in the plane ξ_0, λ_1 stability regions for $R = 0.01$, where the values of other parameters remained constant, equal to $x = \gamma = 1.0$.

It can be seen from the figures that the impact velocity value in the stability regions grows with increasing coefficient of restitution R. However, with increasing R, stability regions become more indented and the indentation appears from the side of negative gaps.

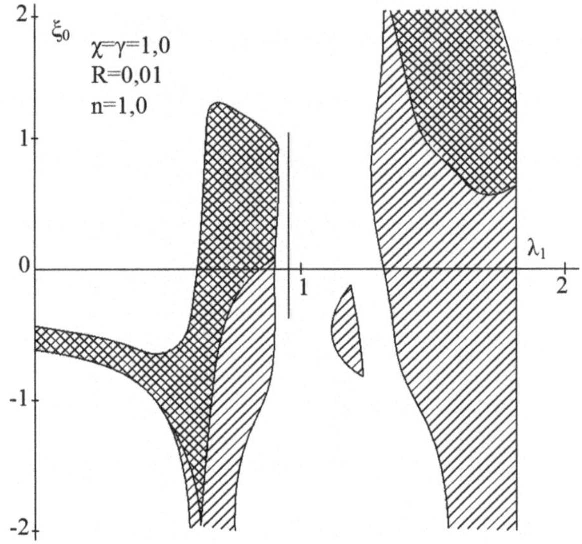

Fig. 2.25 Regions of existence and stability of periodic mode of motion for $\chi = 1, \gamma = 1, R = 0.1, n = 1$

The study of the effect of the n-fold magnitude to the period of external force on the size and location of the stability regions showed that with increasing n, the stability region n arrow significantly over λ_1 and new stability regions emerge.

In the parameter plane ξ_0, λ_1, as was noted above, the two widest stability regions are distinguished near the resonance frequencies of corresponding two-mass linear model. These regions, especially with larger (R.F), are independent of gap change ξ_0; therefore, investigation of the location of these regions with a constant gap as a function of system parameters x and γ is of scientific interest. For simplicity, we assume $\xi_0 = 0$.

Let us consider first the plane λ_1, γ. Figure 2.26 presents stability regions plotted for $x = 1.0$ and $R = 0.2$. The dashed line shows the curves (R.F.) determined from (2.4.20). The stability regions in the vicinity of the largest frequency at changing γ are a series of isolated regions that shift to the left along λ_1 with increasing γ. It should be noted that when one stability region disappears within (R.F), a new one appears which grows in size with increasing γ value and disappears again at some $\gamma_{\kappa p}$, etc. It can be seen that at zero gap in the plane λ_1, γ there is an infinite number of points $\left(\lambda_1^k, \gamma^k\right)$, located at the resonance frequency, for which stable single-impact periodic n-fold motions exist.

When reaching the resonance frequency, stability regions are seen to expand with increasing γ and narrow when withdrawing from it, as shown in Figs. 2.26, and 2.27.

The stability region corresponding to a smaller (R.F) deviates to the right over λ_1 with increasing γ, i.e., to get into this region, it is necessary to decrease the external force frequency ω with growing γ. The stability regions represent a simply connected domain.

In Fig. 2.27, stability regions depending on x are plotted for $\gamma = 1.0$ and $R = 0.2$. Dotted lines correspond to (R.F) curves. The stability regions corresponding

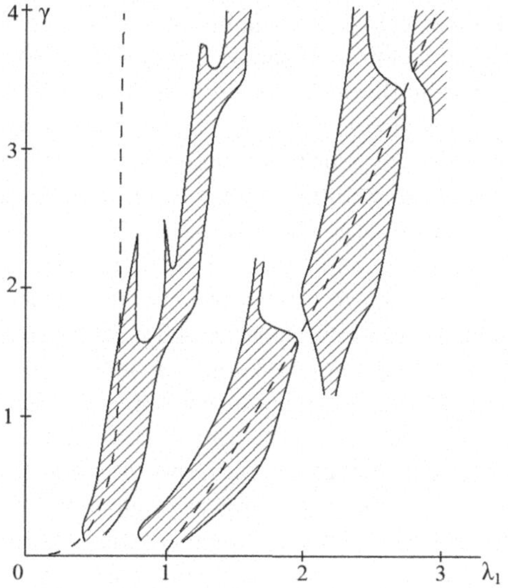

Fig. 2.26 Regions of existence and stability of periodic mode of motion for $\chi = 1, R = 0.2, n = 1$ Resonant frequencies are plotted with dotted lines

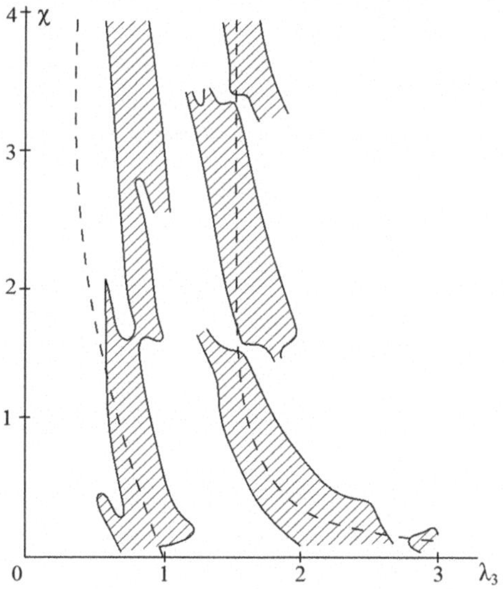

Fig. 2.27 Regions of existence and stability of periodic mode of motion for $\gamma = 1, R = 0.2, n = 1$. Resonant frequencies are plotted with dotted lines

to a higher resonance frequency behave the same way as when $x = 1.0$. The other region, corresponding to a smaller (R.F), shifts to the right over λ_1 at larger x; thus, to get into this region, it is necessary, as in case $x = 1.0$, to reduce the external force frequency.

The dependences shown in Figs. 2.26, and 2.27 are valid up to the coefficient of restitution equal 0.6. For larger R, the stability region corresponding to a higher resonance frequency splits into two, therefore, similar dependencies; the stability region location on parameters can be constructed only for large gap values ξ_0.

Co Study of the location of the stability regions depending on the rate recovery coefficient (see Fig. 2.25) showed that with increasing R, stability regions become more indented from the side of negative gaps, as a result of which similar dependencies of the stability regions location on parameters x and γ for $R > 0.6$ can only be plotted for large gap values.

Let us examine in detail the location of the N-stability regions for the coefficient of restitution $R = 0.05$ and $R = 0.4$ for a different set of parameters x and γ. The value of the coefficient R equal to 0.05 corresponds to the work of the vibro-hammer as a compacting vibro-impact machine, and the value equal to 0.4 corresponds to the work of the vibro-hammer for driving piles, pipes, structures, etc.

Figures 2.28, 2.29, 2.30, 2.31, 2.32 and 2.33 present N-stability regions for one parameter value $x = 0.87$ ($m_1 < m_2$) and for three values γ equal to 10.71; 11.65; and 13.39 for single-impact single and double periodic motions. The coefficient of restitution in this case is either 0.05 or 0.4.

For $\lambda_1 < 2$ domain, N-stability is extremely narrow strips; therefore, plotting of these regions was carried out for $\lambda_1 > 2$. The stability regions of single periodic motions are hatched from the left to the right, and for double motions—from the left down to the right.

As can be seen from the above figures, the stability regions of single-impact periodic motion are rather narrow parts of the plane ξ_0, λ. The region lying in the vicinity of the higher resonance frequency stands out in size among all stability

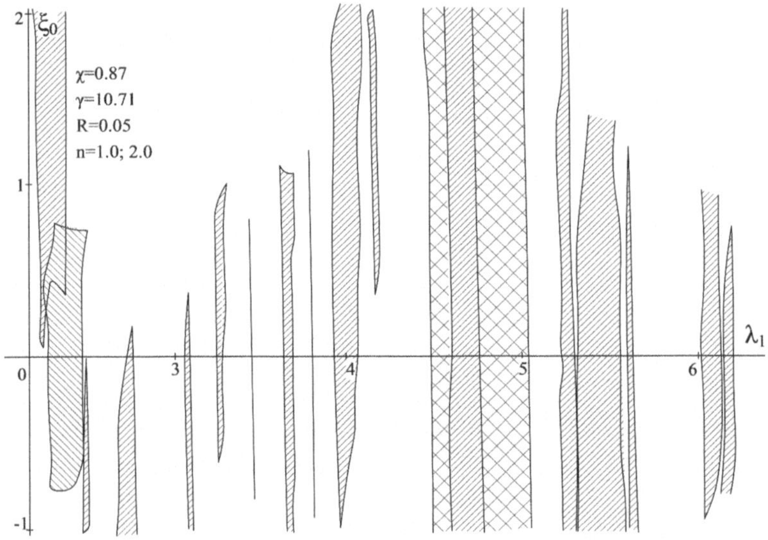

Fig. 2.28 Regions of existence and stability of periodic mode of motion for $\chi = 0.871$, $\gamma = 10.7$, $R = 0.05$, $n = 1$, $n = 2$. Hatching from right down to left for $n = 1$, and from left down to right

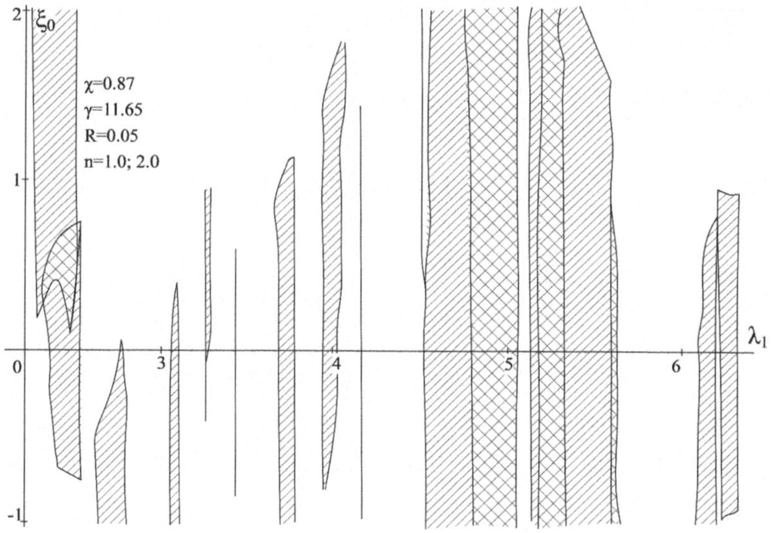

Fig. 2.29 Regions of existence and stability of periodic mode of motion for $\chi = 0.871$, $\gamma = 11.65$, $R = 0.05$, $n = 1$, $n = 2$. Hatching from right down to left for $n = 1$, and from left down to right

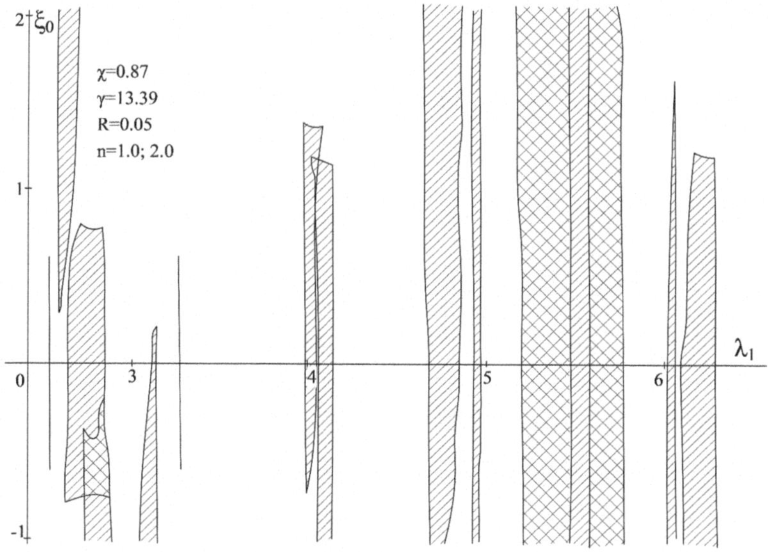

Fig. 2.30 Regions of existence and stability of periodic mode of motion for $\chi = 0.871$, $\gamma = 13.39$, $R = 0.05$, $n = 1$, $n = 2$. Hatching from right down to left for $n = 1$, and from left down to right

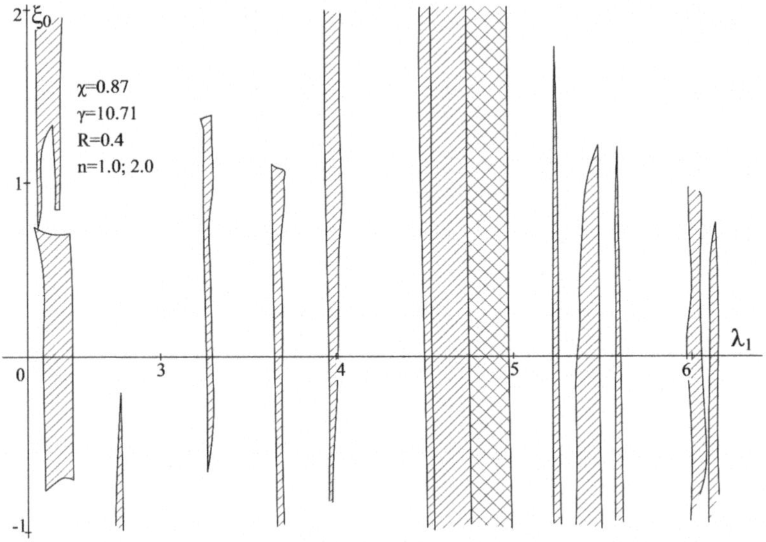

Fig. 2.31 Regions of existence and stability of periodic mode of motion for $\chi = 0.871$, $\gamma = 10.7$, $R = 0.4$, $n = 1$, $n = 2$. Hatching from right down to left for $n = 1$, and from left down to right

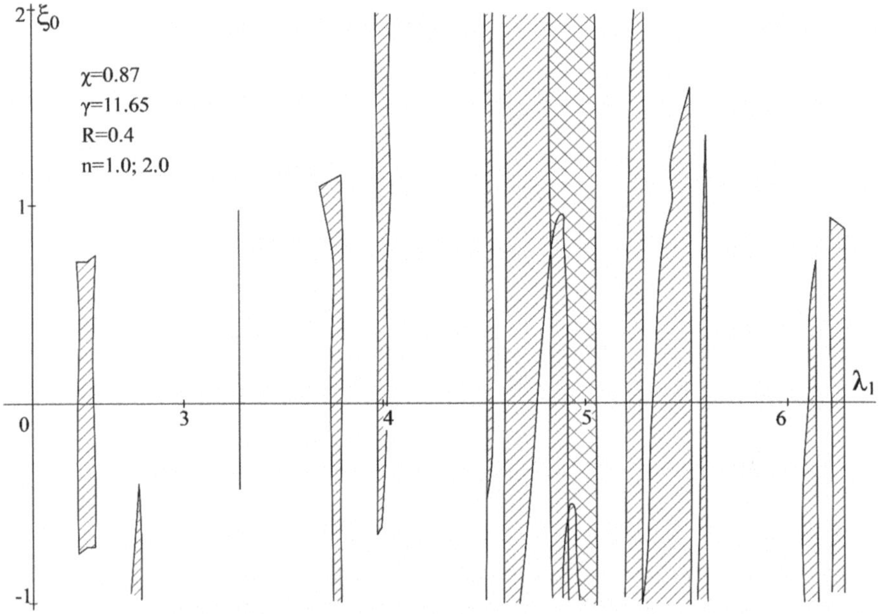

Fig. 2.32 Regions of existence and stability of periodic mode of motion for $\chi = 0.871$, $\gamma = 11.65$, $R = 0.4$, $n = 1$, $n = 2$. Hatching from right down to left for $n = 1$, and from left down to right

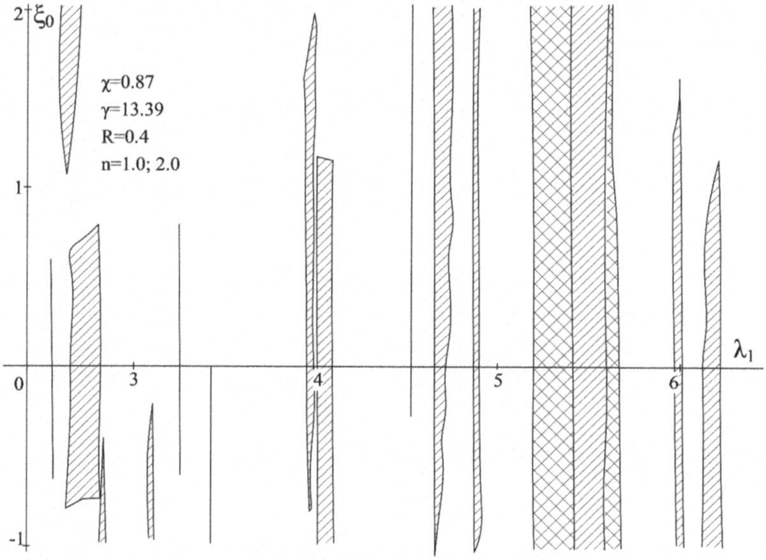

Fig. 2.33 Regions of existence and stability of periodic mode of motion for $\chi = 0.871$, $\gamma = 13.39$ $R = 0.4$, $n = 1$, $n = 2$. Hatching from right down to left for $n = 1$, and from left down to right

regions. The boundaries of this region are located over λ_1 in the range from 4.6 to 5.74.

The change in location of the stability regions depending on the system parameters occurs according to the main foregoing conclusions.

With increasing *n*, the stability regions are getting narrower and their number increases. Due to the narrowness of the stability regions at $n = 2.0$, great practical difficulties arise in maintaining the mechanism in the given mode, although the impact velocities grow approximately by 2.25 times. Therefore, for reliable operation of the mechanism, it is necessary to choose such system parameters that would correspond to its tuning for the simplest single-impact single mode ($n = 1$).

When comparing *N*-stability regions in the parameter plane λ_1, ξ_0 $n = 1$ and $n = 2$, it turns out that two periodic modes can realize in a certain interval λ_1. The existence of two, three, or more stable periodic motion modes has been noted by many researchers (see, e.g., [14, 17, 26, 29, 44] and the bibliography.

With an increase in the rate of the coefficient of restitution upon impact (from 0.05 to 0.4), the stability regions become more indented, and the indentation appears from the side of negative gaps; therefore, to keep the mechanism in the desired stable periodic mode with growing *R*, it is necessary to increase the gap.

In the widest stability regions for various parameter values, the distribution of impact velocities within the region was investigated (Fig. 2.34). In all cases, the maximum impact velocity is observed at the right boundary for negative gaps and for positive gap it shifts to the left boundary.

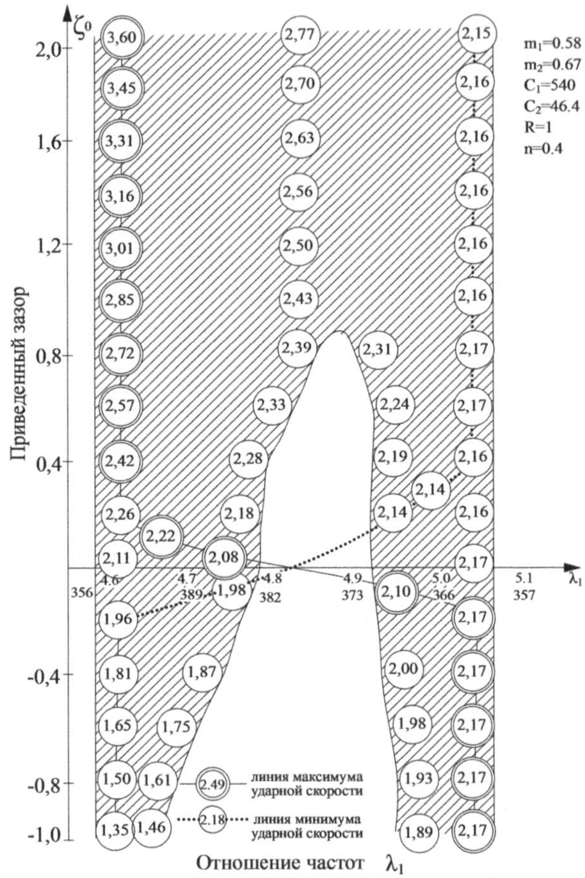

Fig. 2.34 Distribution of impact velocities within stability region

For a value $\dot{\xi}_0$ equal to 5.06, the impact velocity does not change its value ($\dot{\xi}_0 = 2.16$) with gap changing. A practical recommendation in this case is to maintain an external force frequency of 364 rpm. at any gap.

Note that the above theoretical study for $R = 0.05$ and $R = 0.4$ was carried out for parameter values that corresponded to the current self-propelled vibro-impact machine designed at the Central Research Institute of Railways used to compact the crushed stone prism of a railway track [42].

The results of a theoretical study agree well with the experiment. So, if the widest part of the stability region in a theoretical study lies in the range from 4.6 to 5.67, corresponding to ω change in the range from 319 rpm. up to 496 rpm. According to the vibro-impact compactor, ω was selected in the range from 353 rpm. up to 375 rpm. The general distribution of the impact velocity values is in good agreement with the experiment.

2.5 Investigation of the Stability of a Two-Mass Vibro-impact Non-spring-loaded Compactor

In the previous sections, we studied the dynamics of two-mass vibro-impact systems, a characteristic feature of which was that one of the upper or lower masses is elastically connected either to a fixed suspension point or to an oscillation limiter. For full consideration of the general two-mass vibro-impact system (see Fig. 2.1), we still have to examine the system without upper and lower springs.

Investigation of these vibro-impact compactors, both single mass and two mass, is of certain practical interest, since they can be used as self-propelled vibratory compactors and springless vibro hammers.

The dynamics of a single-mass vibro-impact springless mechanism was studied in [3, 5, 45] and others. However, the service life of the systems is significantly reduced due to the fact that the vibrator also serves as an impact mass. Naturally, a special impact mass and a sprung vibrator were introduced. On the one hand, the introduction of a special impact mass and a sprung vibrator increases the service life of the systems, but, on the other hand, the resulted system is a two-mass substantially nonlinear system; the tuning of which to the optimal mode is much more complicated than that of single mass. Therefore, this necessitates the study of the existence and stability regions of a two-mass vibro-impact system, which are essential for choosing the parameter values that correspond to the desired periodic motion.

Below, we consider a vibro-impact system consisting of two masses interconnected by springs with a stiffness coefficient C_2. The lower mass is the impact mass, and the external periodic force $F \sin \omega t$ acts on the upper mass m_2 Fig. 2.35.

The system affects the processed medium via impacts applied by the lower mass over the surface of the processed medium, piles, structures, etc. For tuning and constructing such systems, it is of great interest to single out in the parameters space-specific values at which stable periodic motions exist. In this section, this question is considered in terms of single-impact multiple periodic motions.

When developing a mathematical model, we choose for the reference point a position of the system in which the lower mass m_1 is on the impact surface and the springs are in an undeformed state; the positive direction is indicated by the arrow in Fig. 2.35.

Under the above assumptions, the differential equations of system motion in the intervals between impacts ($x_1 < 0$) for the lower and upper masses, respectively, will have the form

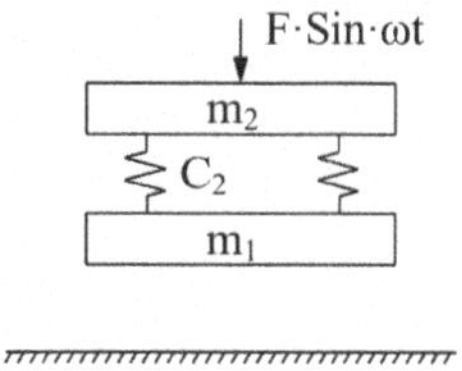

Fig. 2.35 Physical diagram of a non-spring seal

$$
\begin{aligned}
m_1\ddot{x}_1 + C_2(x_1 - x_2) &= Q_1,\\
m_1\ddot{x}_2 + C_2(x_2 - x_1) &= Q_2 + F\sin\omega t.
\end{aligned}
\tag{2.5.1}
$$

and at $x_1 = 0$ an instantaneous elastic impact of the lower mass against the fixed limiter occurs, characterized by the well-known Newton relations

$$
\dot{x}_1^+ = -R\dot{x}_1^-. \tag{2.5.2}
$$

where $\dot{x}_1^+$ and $\dot{x}_1^-$ are lower mass velocities immediately after and before the impact, respectively.

In Eq. (2.5.1), specific mass weight is denoted through Q_1, and Q_2 corresponds to the weight of upper mass, wherein $Q_2 \neq m_2 g$.

In dimensionless variables,

$$
\xi_1 = \frac{m_1\omega^2}{F}x_1,\ \xi_2 = \frac{m_2\omega^2}{F}\left(x_2 - \frac{Q_2}{C_2}\right),\ \omega t = \tau. \tag{2.5.3}
$$

the system (2.5.1) and (2.5.2) will have the following form at $\xi_1 < 0$

$$
\begin{aligned}
\ddot{\xi}_1 + \lambda^2(\xi_1 - x\xi_2) &= q,\\
\ddot{\xi}_2 + \lambda^2(x\xi_2 - \xi_1) &= q + \sin\tau,
\end{aligned}
\tag{2.5.4}
$$

and at $\xi_1 = 0$

$$
\dot{\xi}_1^+ = -R\dot{\xi}_1^-. \tag{2.5.5}
$$

where λ, x, and q denote dimensionless parameters of the form

$$
\lambda^2 = \frac{C_2}{m_1\omega^2},\ x = \frac{m_1}{m_2},\ q = \frac{Q_1 + Q_2}{F}.
$$

Note that the system (2.5.4) explicitly contains a parameter q equal to the ratio of the sum of the mass weights Q_1 and Q_2 to the driving force amplitude, whereas this parameter is not explicitly included in the differential equations of the general two-mass system (2.1.5), (2.1.6). This is explained by the fact that in (2.1.1), in addition to the differences $(x_i - x_j)$ $(i, j = 1, 2, i \neq j)$, there are separate variables x_k $(k = 1, 2)$, allowing to reduce the number of system parameters if desired. On the contrary, differences $(x_i - x_j)$ enter into Eq. (2.5.1), as a result of which the parameter q is an essential parameter of the system, as was observed in the single-mass system.

Similar to the cases considered in §§ 2.1–2.4, it can be established that the image point in the phase space will necessarily repeatedly hit the hyperplane Π ($\xi_1 = 0$). Therefore, the research of the impact motions of the system is reduced to studying the point mapping of the hyperplane Π into itself.

Solving the system of differential Eq. (2.1.16), taking the moment immediately after the impact for the initial conditions, $\tau = \tau_0, \xi_1 = \xi_0 = 0, \dot{\xi}_1 = -R\dot{\xi}_0$, $\xi_2 = \xi_2^0, \dot{\xi}_2 = \dot{\xi}_2^0$, we obtain that the motion of the image point in the space $\xi_1 < 0$ (for $\tau > \tau_0$) is described by the following equations

$$
\begin{aligned}
\xi_1 =& \frac{f_0}{2}\tau^2 + \alpha_1 \sin\tau + C_1 + C_2(\tau - \tau_0) + C_3 \cos\eta(\tau - \tau_0) \\
&+ C_4 \sin\eta(\tau - \tau_0), \\
\dot{\xi}_1 =& f_0\tau + \alpha_1 \cos\tau + C_2 - \eta C_3 \sin\eta(\tau - \tau_0), \\
\xi_2 =& -\frac{q}{x\eta^2}\tau + \frac{f_0}{2x}\tau^2 - \alpha_2 \sin\tau - C_4 \sin\eta(\tau - \tau_0) \\
&+ \frac{1}{x}[C_1 + C_2(\tau - \tau_0) - C_3 \cos\eta(\tau - \tau_0)], \\
\dot{\xi}_2 =& \frac{1}{x}(C_2 + f_0\tau) - \alpha_2 \cos\tau \\
&+ \eta[C_3 \sin\eta(\tau - \tau_0) - C_4 \cos\eta(\tau - \tau_0)].
\end{aligned}
\tag{2.5.6}
$$

where the following notations are introduced

$$
\begin{aligned}
&C_1 = g_0\left(\xi_2^0 + \frac{q}{x\eta^2} - \frac{q}{2}\tau_0^2 + \sin\tau_0\right), f_0 = \frac{xq}{1+x}, \\
&C_2 = g_0\left(\xi_2^0 - \dot{R}\xi_0 - q\tau_0 + \cos\tau_0\right), \\
&C_3 = g_0\left(\xi_2^0 + \frac{q}{x\eta^2} + \frac{\sin\tau_0}{1-\eta^2}\right), \eta^2 = \lambda^2(1+x), \\
&C_4 = \frac{g_0}{\eta}\left(\dot{\xi}_2^0 + \frac{R}{x}\dot{\xi}_0 + \frac{\cos\tau_0}{1-\eta^2}\right), g_0 = \frac{f_0}{q}, \\
&\alpha_1 = \lambda^2 x/\left(1-\eta^2\right), \alpha_2 = \left(1-\lambda^2\right)/\left(1-\eta^2\right).
\end{aligned}
\tag{2.5.7}
$$

We find the moment τ_1 following after τ_0 the image point falls on the hyperplane Π from the first Eq. (2.5.4), assuming $\xi_1 = 0$. From the other three equations, we find, respectively, the values of the lower mass velocity, position, and the upper mass velocity at the moment τ_1 derived from the first equation. The point mapping of a hyperplane Π into itself (let us call it S-transformation) can be written as follows

$$
\begin{aligned}
-C_1 =& \frac{f_0}{2}\tau_1^2 + \alpha_1 \sin\tau_1 + C_2(\tau_1 - \tau_0) + C_3 \cos\eta(\tau_1 - \tau_0) \\
&+ C_4 \sin\eta(\tau_1 - \tau_0), \\
\dot{\xi}_1 =& f_0\tau_1 + \alpha_1 \cos\tau_1 + C_2 \\
&- \eta[C_3 \sin\eta(\tau_1 - \tau_0) - C_4 \cos\eta(\tau_1 - \tau_0)], \\
\xi_2 =& -\frac{q}{x\eta^2}\tau + \frac{f_0}{2x}\tau_1^2 - \alpha_2 \sin\tau_1
\end{aligned}
$$

$$
\begin{aligned}
&+\frac{1}{x}[C_1 + C_2(\tau - \tau_0) - C_3 \cos \eta(\tau_1 - \tau_0)] \\
&- C_4 \sin \eta(\tau_1 - \tau_0), \\
\dot{\xi}_2 &= \frac{1}{x}(C_2 + f_0\tau_1) - \alpha_2 \cos \tau_1 \\
&+ \eta[C_3 \sin \eta(\tau_1 - \tau_0) - C_4 \cos \eta(\tau_1 - \tau_0)]. \qquad (2.5.8)
\end{aligned}
$$

Note that of all the roots τ_1 of the first Eq. (2.5.8), it is necessary to choose the smallest one satisfying the condition $\tau_1 > \tau_0$.

It follows from (2.5.8) that any point of the hyperplane Π $A_0\left(\tau_0, \dot{\xi}_1^0, \xi_2^0, \dot{\xi}_2^0\right)$ is transformed to another point of the same hyperplane $A_1\left(\tau_1, \dot{\xi}_1^1, \xi_2^1, \dot{\xi}_2^1\right)$ with the help of S-transformation.

We proceed with studying periodic modes. The case of m-impact n-fold periodic motion corresponds to a fixed-point S^m-mapping with periodicity conditions which can be written in the following form

$$
\dot{\xi}_1^m = \dot{\xi}_1^0, \xi_2^m = \xi_2^0, \dot{\xi}_2^m = \dot{\xi}_2^0, \tau_m - \tau_0 = T(n, m = 1, 2 \ldots). \qquad (2.5.9)
$$

The selection of the smallest τ roots of all the roots satisfying S-mapping can be written in the form that the displacement of the impact mass m_1 should not be positive for $\tau_0 \leq \tau \leq \tau_m$, i.e.,

$$
\xi_1(\tau) \leq 0, \text{ for all } \tau_0 \leq \tau \leq \tau_m. \qquad (2.5.10)
$$

Writing out S^m-mapping formulas and using relation (2.5.9) to determine the coordinates of the invariant point S-mapping $\dot{\xi}_1^0$, ξ_2^0, $\dot{\xi}_2^0$ and τ_0 we will derive the following system $4(m+1)$ of equations

$$
\begin{aligned}
0 &= \frac{f_0}{2}\tau_i^2 + \alpha_1 \sin \tau_i + C_1 + C_2(\tau_i - \tau_{i-1}) + C_3 \cos \eta(\tau_i - \tau_{i-1}) \\
&+ C_4 \sin \eta(\tau_i - \tau_{i-1}), \\
\dot{\xi}_1^i &= f_0\tau_i + \alpha_1 \cos \tau_i + C_2 \\
&- \eta\left[C_3 \sin \eta(\tau_i - \tau_{i-1}) - C_4 \cos \eta(\tau_i - \tau_{i-1})\right], \\
\xi_2^i &= -\frac{q}{x\eta^2} + \frac{f_0}{2x}\tau_i^2 - \alpha_2 \sin \tau_i \\
&+ \frac{1}{x}\left[C_1 + C_2(\tau_i - \tau_{i-1}) - C_3 \cos \eta(\tau_i - \tau_{i-1})\right] \\
&- C_4 \sin \eta(\tau_i - \tau_{i-1}), \\
\dot{\xi}_2^i &= \frac{1}{x}(C_2 + f_0\tau_i) - \alpha_2 \cos \tau_i \\
&+ \eta\left[C_3 \sin \eta(\tau_i - \tau_{i-1}) - C_4 \cos \eta(\tau_i - \tau_{i-1})\right], \\
\dot{\xi}_1^m &= \dot{\xi}_1^0, \xi_2^m = \xi_2^0, \dot{\xi}_2^m = \dot{\xi}_2^0, \tau_m - \tau_0 = 2\pi n,
\end{aligned}
$$

$$i = 1, 2, \ldots, m. \tag{2.5.11}$$

Considering further only single-impact ($m = 1$) n-fold periodic motions, the system (2.5.11) for finding the coordinates of a fixed-point S-mapping can be written in the form

$$\begin{aligned}
0 &= \frac{f_0}{2}(\tau_0 + T)^2 + \alpha_1 \sin \tau_0 + C_1 + C_2 T + C_3 \cos \eta(\tau_i - \tau_{i-1}) \\
&\quad + C_4 \sin \eta T, \\
\dot{\xi}_1^0 &= f_0(\tau_0 + T) + \alpha_1 \cos \tau_0 + C_2 - \eta[C_3 \sin \eta T - C_4 \cos \eta T], \\
\xi_2^0 &= -\frac{q}{x\eta^2} + \frac{f_0}{2x}(\tau_0 + T)^2 - \alpha_2 \sin \tau_0 \\
&\quad + \frac{1}{x}[C_1 + C_2 T - C_3 \cos \eta T] - C_4 \sin \eta T, \\
\dot{\xi}_2^0 &= \frac{1}{x}\big[C_2 + f_0(\tau_0 + T)\big] - \alpha_2 \cos \tau_0 \\
&\quad + \eta[C_3 \sin \eta T - C_4 \cos \eta T].
\end{aligned} \tag{2.5.12}$$

From the first and third Eq. (2.5.12), we deduce that

$$\begin{aligned}
\xi_2^0 &= -\frac{q}{x\eta^2} + \frac{f_0}{2x}(\tau_0 + T)^2 \frac{1 + x}{x} + (\alpha_1 - \alpha_2) \sin \tau_0 \\
&\quad + \frac{1 + x}{x}[C_1 + C_2 T].
\end{aligned} \tag{2.5.13}$$

And for the second and fourth equations of the same system, we have

$$\dot{\xi}_1^0 = \dot{\xi}_2^0 + \big[C_2 + f_0(\tau_0 + T)\big]\frac{1 + x}{x} + (\alpha_1 - \alpha_2) \cos \tau_0. \tag{2.5.14}$$

Finally, substituting (2.5.13, 2.5.14) into the third and fourth Eq. (2.5.12), we find expression for the coordinates of a fixed point in parametric form, corresponding to a single-impact periodic mode of S-mapping

$$\begin{aligned}
\dot{\xi}_1^0 &= \frac{qT}{(1 + R)}, \quad \dot{\xi}_2^0 = \frac{2(1 - R)(\lambda^2 - 1)}{x\lambda^2}\dot{\xi}_1^0, \\
\cos \tau_0 &= \frac{2(1 - \eta^2)(1 - R)}{x\lambda^2}, \\
\xi_2^0 &= \frac{qT}{2x\eta} ctg(\pi n \eta) - \frac{\sin \tau_0}{1 - \eta^2} - \frac{q}{x\eta^2}.
\end{aligned} \tag{2.5.16}$$

It follows from the relation (2.5.16) that for fixed values of λ, x, q and R parameters there exist two fixed mapping points A_1^* and A_2^* with the following coordinates

$$A_1^*\left(\dot{\xi}_1^0, \xi_2^0, \dot{\xi}_2^0, \arccos a\right)$$
$$A_2^*\left(\dot{\xi}_1^0, \xi_2^0, \dot{\xi}_2^0, 2\pi - \arccos a\right)$$

where $a = 2\left(1 - \eta^2\right)(1 - R)/x\lambda^2$.

However, as will be shown below, at $\eta < 1$ the fixed point A_2^* is unstable, and at $\eta > 1$, A_1^* is unstable.

For the existence of a fixed point, the condition $|\cos \tau_0| \le 1$ is to be imposed, which has the form

$$-x\eta^2(1 + R) \le \left(1 - \eta^2\right)\pi nq(1 + x)(1 - R) \le (1 + R)x\eta^2. \tag{2.5.17}$$

The right inequality provides one boundary of existence

$$\lambda^2\left[\pi nq(1 - R)(1 + x) + x(1 + R)\right] \ge \pi nq(1 - R). \tag{2.5.18}$$

And it is derived from the left inequality that at $\pi nq(1 + x)(1 - R) - x(1 + R) > 0$ the relation is held

$$\pi nq(1 - R) \ge \lambda^2\left[\pi nq(1 + x)(1 - R) - (1 + R)x\right]. \tag{2.5.19}$$

and at $\pi nq(1 + x)(1 - R) - x(1 + R) < 0$ the left inequality does not provide boundary of existence.

Inequalities (2.5.18), (2.5.19) determine the lower and upper boundary of the existence region in the parameter plane λ, x, respectively. It follows from (2.5.18), (2.5.19) that at

$$x \to \frac{(1 - R)\pi nq}{(1 + R) - \pi nq(1 - R)}$$

the upper boundary of existence region goes to infinity ($\lambda \to \infty$), and at $R \to 1$, $x \to 0$ the lower boundary approaches the abscissa axis. Hence, the widest region of existence of periodic mode can be achieved by using the following approximate formulas for a combination of parameters

$$q < (\pi n)^{-1}, \frac{\pi nq - 1}{\pi nq + 1} < R. \tag{2.5.20}$$

We note that impact velocity (2.5.16), as in case of a single-mass springless vibro-striker [18] is maximal at absolutely inelastic impact.

It follows from (2.5.18), (2.5.19) that at $\lambda, x \to \infty$ (which corresponds to the transition from a two-mass model to a single-mass springless model) the expressions for the coordinates of the fixed point coincide up to the notations with the corresponding formulas in [45].

The region of existence of a single-impact periodic regime determined by inequalities (2.5.18), (2.5.19) can include motions with transition beyond the boundary plane, since when finding the coordinates of a fixed point, the condition $\xi_1 \leq 0$ for all $\tau_0 \leq \tau \leq \tau_1 (\tau_1 - \tau_0 = T)$ was not employed anywhere. The case of the motion trajectory touching the plane of the limiter is boundary condition. These bifurcation boundaries that correspond to breaking points of S-mapping (or to additional impacts for the motion period) are denoted through C_τ and determined by the following relations

$$\begin{aligned}
&\xi_1(\tau^*) = 0, \dot{\xi}_1(\tau^*) = 0, \tau_0 \leq \tau^* \leq \tau_1, \\
&\quad 0 = \frac{f_0}{2}\tau^{*2} + \alpha_1 \sin \tau^* - C_1 + C_2\beta + C_3 \cos \eta\beta + C_4 \sin \eta\beta, \\
&\quad 0 = f_0\tau^* + \alpha_1 \cos \tau^* + C_2 - \eta[C_3 \sin \eta\beta - C_4 \cos \eta\beta].
\end{aligned} \tag{2.5.21}$$

where $\beta = \tau^* - \tau_0$.

From system (2.5.21) using relations (2.5.16), after some transformations we obtain

$$\begin{aligned}
tg\frac{\beta}{2} &= \Phi^{-1}\left[\frac{1 - \cos \eta\beta}{\eta} ctg(\pi n q) - x\beta \right. \\
&\quad \left. + \frac{\sin \eta\beta}{\eta} + \frac{x\beta^2}{T} + \frac{1-R}{1+R}(1+x)\sin\beta\right], \\
q &= \frac{x\eta^2|\sin\beta|}{|1-\eta^2|\pi n\left\{\Phi^2 + \left[(1-R)(1+R)^{-1}(1+x)\sin\beta\right]^2\right\}^{1/2}}.
\end{aligned} \tag{2.5.22}$$

where Φ denotes

$$\Phi = -x + \sin \eta\beta ctg(\pi n q) - \cos \eta\beta + \frac{2x\beta}{T} + \frac{1-R}{1+R}(1+x)\cos\beta. \tag{2.5.23}$$

From the first Eq. (2.5.22) at fixed parameters R, x, and η we find β, and from the second equation, we define q by R, x, η, and β.

To study the stability of a fixed point of a point mapping S, and therefore the corresponding periodic motion, we set up the characteristic equation $\chi(z) = 0$.

After linearization of the point mapping S, a homogeneous system of linear equations with respect to variations is obtained. For the existence of a nontrivial solution, it is necessary and sufficient that the determinant of the system

$$\begin{vmatrix} 0 & 1-z & T & (z+R)\dot{\xi}_1^0 + (z-1)\dot{\xi}_2^0 \\ z(1+z) & 0 & z-1 & 0 \\ -\eta R \sin T\eta & x\lambda^2(1-\cos\eta T) & x\lambda^2\left(1-\frac{\sin\eta T}{\eta}\right) & z\dot{\xi}_1^0\eta^2 - \eta F\sin\eta T + \lambda^2\left[\left(x\dot{\xi}_2^0 + R\dot{\xi}_1^0\right)\cos\eta T + x\left(R\dot{\xi}_1^0 - \dot{\xi}_2^0\right)\right] \\ \eta^2(z+R\cos\eta T) & -x\eta\lambda^2\sin\eta T & x\lambda^2(\cos\eta T - 1) & \eta^2 F(\cos\eta T - z) + \eta\lambda^2\left(x\dot{\xi}_2^0 + R\dot{\xi}_1^0\sin\eta T\right) \end{vmatrix} \tag{2.5.24}$$

equals to zero. In (2.5.24), F denotes $q + x\lambda^2\xi_2^0$. Expanding the determinant and equating it to zero, we obtain the characteristic equation

$$\chi(z) = \sum_{i=0}^{4} A_i z^{4-i} = 0 \tag{2.5.25}$$

which can be written in the form

$$\begin{aligned}(1+x)\eta\dot{\xi}_1^0\left(z^4+R^2\right) + \{2\left[(F_1+F_2c)\left(R-z^2\right)\right] \\ +F_3(-1-F_4)\left(z^2+1\right)\}z \\ +\{2F_3(1+F_4c) + (1+x)^{-1} \\ \left[F_1^2+F_2^2+F_5+F_4(1+R)^2\rho\right]\}z^2 = 0\end{aligned} \tag{2.5.26}$$

where the following notation are introduced

$$\begin{aligned} &F_1 = (x-R)\dot{\xi}_1^0,\ F_2 = (1-xR)\eta\dot{\xi}_1^0,\ c = \cos\eta T \\ &F_3 = (1+R)F,\ F_4 = T\eta x,\ \rho = \sin\eta T \\ &F_5 = \left[2x(1-R)^2 - 4R\left(1+x^2\right)\right]\eta\dot{\xi}_1^0 c \end{aligned} \tag{2.5.27}$$

To determine the stability of the obtained fixed point of the point mapping, we can use the Routh–Hurwitz inequalities for the polynomial coefficients (2.5.26), transformed using the formulas (2.3.12). Hence, we obtain additional relations for phase τ_0. This is due to the fact that in the phase space of the system with the same parameter values there is not a periodic motion of the specified type or there are two at the same time, i.e., there are two fixed points of the point mapping S.

Let us show that at $\eta > 1$, the fixed point A_1^* is always unstable and at $\eta < 1$, A_2^* is always unstable.

It is known that for the stability of a fixed point the following inequality is to be held (at $A_0 > 0$)

$$\sum_{i=0}^{4} A_i > 0 \tag{2.5.28}$$

which, with account for (2.5.26), can be written in the form

$$(1+x)\eta\dot{\xi}_1^0(z^4+R^2)+\{2[(F_1+F_2c)(R-1)]+F_3(1+F_4)\} \\ +\{2F_3(1+F_4c)+(1+x)^{-1}[F_1^2+F_2^2+F_5+F_4(1+R)^2\rho]\}>0 \tag{2.5.29}$$

By substituting (2.5.27), (2.5.16) into (2.5.29), we obtain another additional condition for determining the phase τ_0, namely (at $\eta \neq 1$)

$$(1-\eta^2)\sin\tau_0 > 0(\eta = \lambda\sqrt{1+x}) \tag{2.5.30}$$

It immediately follows from inequality (2.5.30) that at $\eta < 1$, $\tau_0 \in [0, \pi]$, since $\sin\tau_0 > 0$, and at $\eta < 1$ $\tau_0 \in [\pi, 2\pi]$. This means that at $\lambda\sqrt{1+x} < 1$ unstable periodic motion corresponds to a fixed point A_2^*, and at $\lambda\sqrt{1+x} > 1$ it corresponds to A_1^*. We note that the curve

$$\lambda = (1+x)^{-1/2} \tag{2.5.31}$$

dividing the parameter plane x, λ into two half-planes: the first $-\eta^2 < 1$ and the second $-\eta^2 > 1$ in which the periodic motion is stable only in either A_1^* or A_2^* fixed points is a resonance curve for a linear two-mass model without a limiter. We call the curve (2.5.31) for brevity R.F.

Let us consider some plane sections of the 4-D parameter space, in which we will partition the plane of two parameters into stability regions.

In plane x, λ for fixed values of the restitution coefficient R and parameter $q = \frac{Q_1+Q_2}{F}$ the stability regions are rather narrow strips extending from the boundary of existence (2.5.18), (2.5.19) and with increasing x and λ these regions are getting n arrower. This is easily explained from a physical point of view, as at $x \to \infty$ (which corresponds to an increase in the impact mass m_1 and a decrease in the vibrator mass) the driving force $F\sin\omega t$ cannot impose on the system the periodic motion due to the infinite mass of the striker (it simply will not separate it from the limiter). At $\lambda \to \infty$ that corresponds to:

1. the frequency of the driving force ω tending to zero,
2. mass of the striker tending to zero.

In the first case, the system cannot have periodic modes as it is not affected by a sinusoidal external force ($\omega = 0$). In the second case, impact motions in the system cannot exist since the mass of the striker is zero and the vibrator, by assumption (A), does not experience impacts against the limiting plane.

Figure 2.36 presents for $R = 0.95$ and $q = 1$, the partitioning of the parameter plane x, λ into stability regions (the shaded part of the plane) for a single-impact periodic motion of $n = 1$ and 2 multiplicity. It can be seen from the figure that the widest part

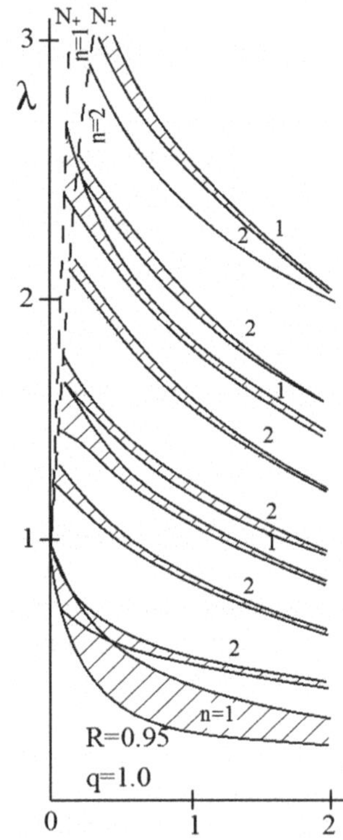

Fig. 2.36 Periodic regime stability regions for $R = 0.95$, $q = 1$

of the stability region lies for values $\lambda < 1$, i.e., with such a ratio of parameters when the frequency of the driving force is higher than the eigen partial frequency of the system. A numerical calculation of the x, λ plane partition into the stability region showed that, with increasing R and q values, the stability regions expand, with its lower boundary approaching the straight line $\lambda = 0$, and the upper boundary tends to the straight line $x = 0$.

For values $\lambda > 1$, the stability regions are narrow stripes and due to it, the stable regime is possible only for small values x ($x < 1$).

The studies shown that with decreasing the coefficient of restitution R the stability regions are getting so narrow in the plane x, λ that they are practically disappear in it.

As expected (see §§ 2.1–2.4), the stability regions get narrow significantly with increasing number of multiplicities of the period of oscillations to the period the driving force. An overlap of stability regions is observed for different values n. This suggests that in this system, periodic motions with different multiplicities are possible depending on the initial conditions at the same parameter values.

When constructing a bifurcation curve C_τ, it turned out that it intersects nothing from the stability region.

Let us consider the plane of parameters R, q. Figure 2.37 presents the stability regions (the shaded part of the plane) for $x = 0.12$ in the plane R, q at $\lambda = 0.8$ for

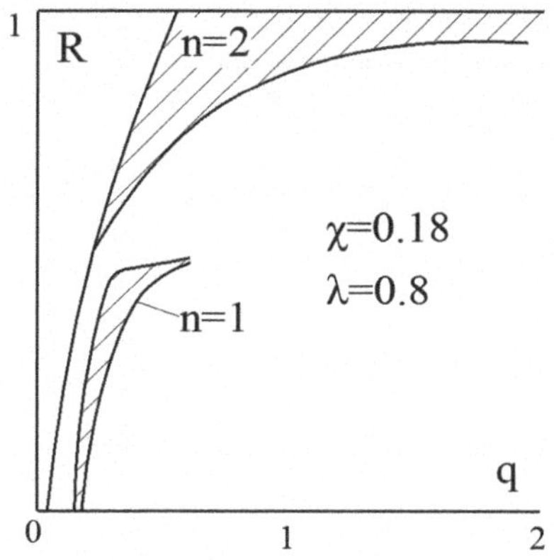

Fig. 2.37 Stability areas of periodic motion mode

$n = 1, 2$. It can be seen from the figure that position of the stability regions in this parameter plane is similar to that for the case of a single-mass springless vibration striker [45].

At fixed values $\lambda = \lambda_0$, the stability regions for $n = 1$ are getting wider with increasing x till the value $x = x^*$. At $x > x^*$, the stability region forth is mode is seen to disappear but further, with increasing x, it appears again but exists for a smaller interval over x and then disappears, etc. Thus, in the cross-section $\lambda = \lambda_0$ there exists an infinite number of segments over x $\{\alpha_i, \beta_i\}$ with stable periodic motion, and the length of these segments tends to zero, in other words the following relation holds.

$$\frac{\beta_{i+1} - \alpha_i}{\beta_i - \alpha_i} < 1, \ \lim_{i \to \infty} \beta_i - \alpha_i = 0. \tag{2.5.32}$$

The widest range over x is found for the values $\lambda < 1$, i.e., for those values when the frequency of the driving force is greater than the eigen frequency of the system. Remind that the parameter region of the system in which the periodic motion is stable corresponds to a fixed point A_1 and the motion corresponding to a fixed point A_2^* is always unstable. We also note that this stability region is near the resonance curve of the corresponding two-mass model, the equation of which has the form (2.5.31).

As a result of the calculation, it turned out that with increasing λ the interval of the stability regions with respect to x in the plane R, q decreases. Thus, at $\lambda = 0.1$ the stability region exists for $x \in [0.1; 25.2]$, whereas for $\lambda = 0.8$ it already exists in the interval $[0; 0.2]$.

The loss of stability for the mode $n = 1$ for some x^* does not always entail the loss of stability for the modes with $n = 2, 3, \ldots$. This is explained by the fact that when growing, the stability regions in the parameter plane x, λ are divided into smaller ones and herein new stability regions appear. Therefore, at a fixed value $\lambda = \lambda_0$ with increasing x, this straight line $\lambda = \lambda_0$ intersects stability regions either in the part common for $n = 1, 2, 3 \ldots$, or in the region where the mode is stable only at $n = 1$ or $n = 2$ $(3, 4 \ldots)$. Figure 2.38 shows the loss of stability mode for $n = 2$, and in

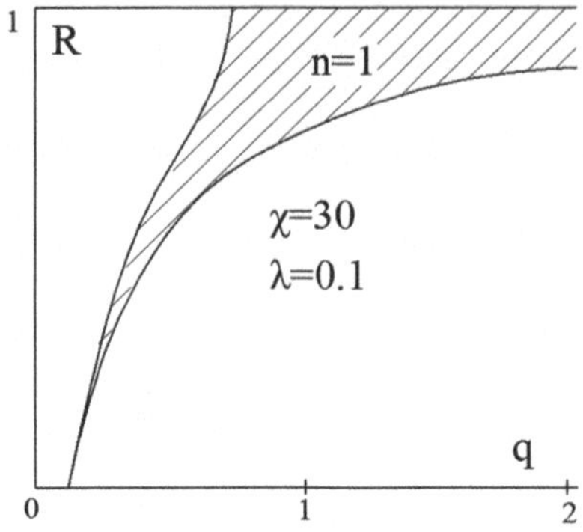

Fig. 2.38 Regions of stability of periodic motion

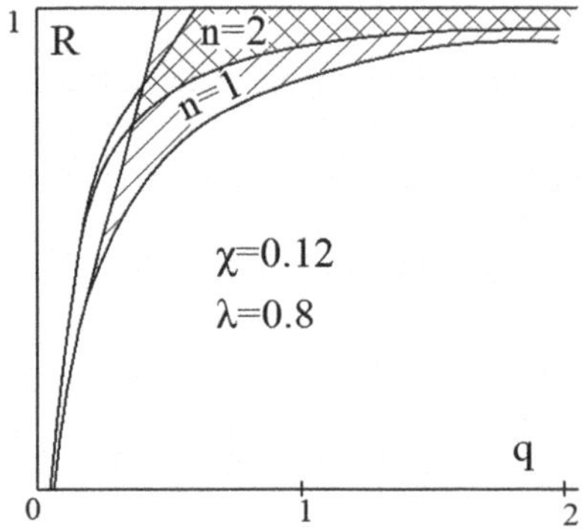

Fig. 2.39 Stability areas of periodic motion mode

Fig. 2.39, the stable mode for $n = 1$ disappears at $x = 30$, $\lambda = 0.1$ and $x = 0.18$, $\lambda = 0.8$, respectively.

Figure 2.40 shows the change in position of the stability regions in R, q plane for $\lambda = 0.3$ with varying x. One can clearly see that stability modes disappear and appear again either at $n = 2$, or at $n = 1$. At $x = 0.8$, the mode with $n = 2$ practically disappears due to the fact that the stability boundaries of this mode merge with the straight line $q = 0$. With increasing x, the stability region of the mode for $n = 1$ expands up to $x = 2$, and then at $x = 4$, it disappears. At $x = 22$, the region appears again and exists up to $x = 26$, etc.

When constructing bifurcation curves in this plane, it turned out that they cut off a small part from the stability region, moreover, at small R. This result coincides with such conclusions in [45], with the only exception that in our case the stability regions are much narrower than small R.

Thus, from the results obtained above, it follows that the widest stability regions are observed when the coefficient of restitution upon impact tends to one. Therefore, we consider this limiting case of an absolutely elastic impact ($R = 1$).

Before analyzing the dynamics of the system with absolutely elastic impacts, we note some general properties of vibro-impact systems with absolutely elastic impacts.

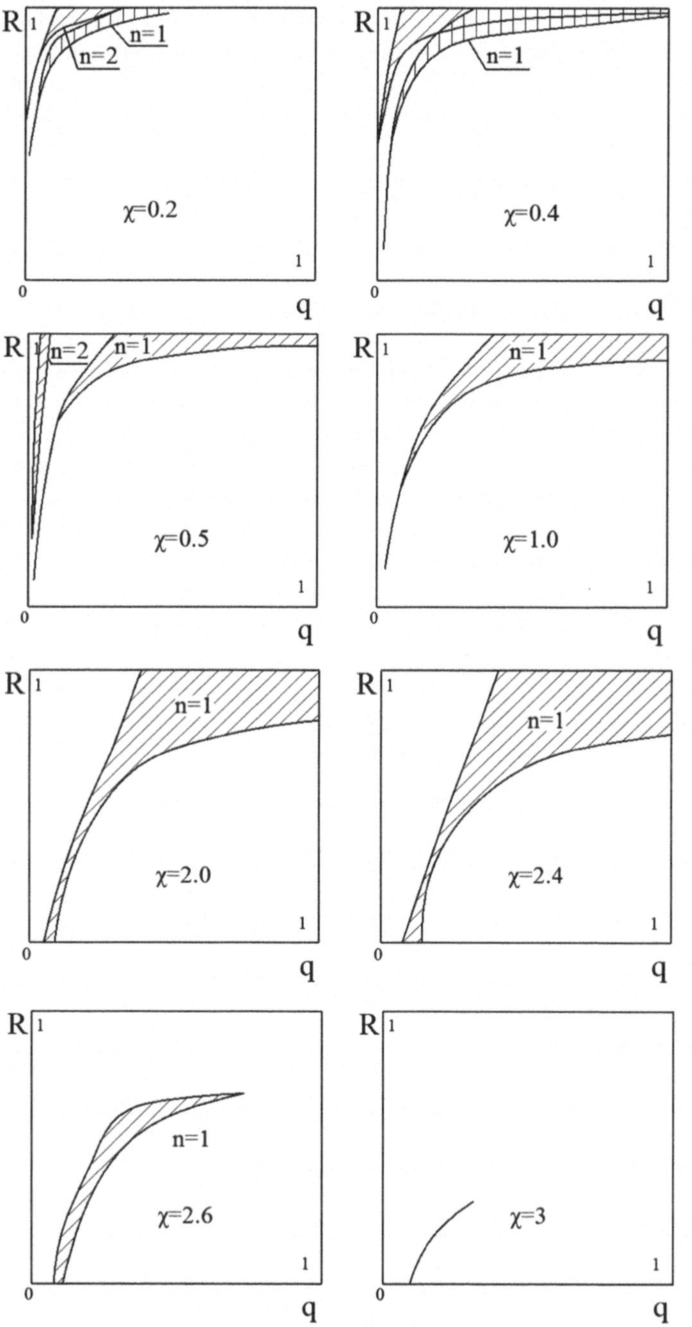

Fig. 2.40 Stability areas of periodic motion modes for various values χ

First, we note that surface $R = 1$ in 4-D parameter space λ, x, R and q in our case (or in N-D parameter space of any other vibro-impact system) corresponds to the boundary of the stability region N_φ. Secondly, the surfaces corresponding to other roots of the characteristic equation (for instance $Z = 1$ or $Z = -1$ and together with $Z = \pm 1$) can intersect with the boundary surface N_φ with respect to some surface configurations. Therefore, the problem of studying the stability in the vicinity of this boundary involves searching the "lines" (points in the case of a plane with two parameters) that correspond to the case when other roots of the characteristic polynomial lie in the unit circle. It is often difficult to determine the location of the stability region by N-surface equations. For this, it is necessary to find the points of intersection of the stability region boundary with other (if any) boundary curves N_+ and N_-. After finding these intersection points, called parametric points [46], it is necessary to determine their multiplicity using the conditions found in the work [46].

After identifying the multiplicity of the parametric point, it is necessary to determine the motion direction along the curve N_+ (N_-) depending on the change in the system parameters. For this, there are inequalities in [46] that provide a transition to a region with a large number of roots inside the unit circle.

Thus, for numerical construction of the stability region, it is necessary to know the starting points for the surfaces on the boundary N_φ, from which the construction of these boundary stability surfaces begins. By continuing numerically these boundaries in the parameter space, one can thus obtain the entire stability region.

We proceed analyzing the case of absolutely elastic impact. Coordinates of a fixed point of S-mapping in this case are written in the form

$$\dot{\xi}_1^0 = \frac{q}{2}T, \cos\tau_0 = 0,$$
$$\xi_2^0 = \frac{q}{2}T(x\eta)^{-1}ctg\left(\frac{T}{2}\eta\right) - \frac{q}{|1-\eta^2|} - \frac{q}{x\eta^2}. \tag{2.5.33}$$

It is evident from (2.5.33) that existence regions of the periodic mode, the boundary of which coincides with the stability boundary N_+, represent in 3-D parameter space x, λ, and q, a quadrant $x \geq 0$, $\lambda \geq 0$ и $q \geq 0$. Hence, the stability region can be limited by N_-- and N_φ-surfaces, since the boundary N_+, coinciding with the boundary of existence, does not provide the boundary of the stability region.

To determine the stability of a fixed point (2.5.33) S-mapping, we compose a characteristic polynomial, which can be written in the following form

$$\begin{aligned}\lambda^2(1+x)^2\dot{\xi}_1^0(z^4+1) + 2\{\lambda^2(1-x)^2(c-1)\dot{\xi}_1^0 \\ -\eta(q+x\lambda^2\xi_2^0)(x\eta T+1)\} * z(z^2+1) \\ +2\{\lambda^2[(1-x)^2c] + 2x\eta T\rho)\dot{\xi}_1^0 \\ +2\eta(q+x\lambda^2\xi_2^0)(1+x\eta Tc)\} * z^2 = 0.\end{aligned} \tag{2.5.34}$$

The loss of stability of a single-impact x-multiple periodic mode occurs when crossing the boundary N_+, the equation of which is written in the form

$$|\cos\tau_0| = 1. \quad (2.5.35)$$

The stability violation with equality $x = -1$ occurs when crossing the boundary N_+, the equation of which has the form

$$1 + \frac{T}{2}x\eta ctg\left(\frac{T}{2}\eta\right) = 0. \quad (2.5.36)$$

$$q[\eta T + x\sin T\eta] = \frac{x\eta^2 \sin T\eta}{|1-\eta^2|}. \quad (2.5.37)$$

To obtain the boundary N_φ, one cannot proceed from the characteristic polynomial (2.5.34), obtained from the general polynomial (2.5.26), assuming $R = 1$, since the part of the hyperplane $R = 1$, located between N_+ and N_-, is the stability boundary N_φ.

In order to obtain the boundary N_φ, we proceed as follows: for the fourth-degree characteristic Eq. (2.5.26), the expression for the boundary N_φ is written in terms of the coefficients of the polynomial as follows

$$(A_0 - A_4)^2(A_0 + A_4 - A_2) + (A_3 - A_1)(A_1A_4 - A_0A_3) = 0. \quad (2.5.38)$$

Substituting A_i ($i = 0, 1, 2, 3, 4$) coefficients of the characteristic polynomial (2.5.26) into equality (2.5.38), we obtain, after some transformations, an equation determining the boundary N_φ in the form

$$\left(1-R^2\right)^2\left\{\begin{array}{l}\eta\dot{\xi}_1^0\left[\left(1-R^2\right)(1-c)+2R\rho^2-\frac{T}{2}\eta(1+R)^2\rho\right]\\ +(1+R)\left(q-x\lambda^2\xi_2^0\right)(1-c)(T\eta-\rho)\end{array}\right\}=0. \quad (2.5.39)$$

Having obtained an explicit expression of boundary N_φ in terms of the system parameters, we thereby showed that the surface $R = 1$ is the boundary of the stability region. It can also be seen from (2.5.39) that, for $R = 1$, it becomes null equation. Therefore, if we use the characteristic Eq. (2.5.34) to obtain the equations defining the boundary N_φ, then it turns out that the boundary N_φ does not exist, as the coefficients of this polynomial were obtained from the coefficients (2.5.26) the condition $R = 1$ having been already used. Thus, to find equations for the boundary N_φ in the case of absolutely elastic impact, the simplified coefficients of the characteristic Eq. (2.5.34) obtained from (2.5.26) by substituting them with$R = 1$ cannot be applied.

$$A_i^{`} = A_i(R = 1)(i = 1, 2, 3, 4).$$

Note that in the general case the expression for the boundary N_φ is the product of two functions Φ_1 and Φ_2, one of which has the following property (suppose Φ_1).

$$\Phi_1(R=1)=0,\ \Phi_1(R=0)=1.$$

Function Φ_2 has the property $\Phi_2(R=1)\neq 0$. In our case, the equation $\Phi_2(R=1)$ has the form

$$\Phi_2(R=1)=2(\rho-T\eta)\left\{\eta\dot{\xi}_1^0\rho+q+x\lambda^2\right.$$
$$\left[\frac{q}{2}T(x\eta)^{-1}ctg\left(\frac{T}{2}\eta\right)-\frac{1}{|1-\eta^2|}-\frac{q}{x\eta^2}\right](1-c) \tag{2.5.40}$$

which is identical but not equal to zero.

Thus, the equation for determining the boundary N_φ splits in the space of three parameters x, η, and q into two equations

$$\eta=0,\ q=\frac{\eta^2 tg\frac{T}{2}\eta}{|1-\eta^2|\left(tg\left(\frac{T}{2}\eta\right)-\frac{T}{2}\eta\right)} \tag{2.5.41}$$

It can be seen from this expression that the boundary N_φ in 3-D parameter space x, η, and q is a plane $\eta=0$ and a cylindrical surface with a generatrix parallel to the axis x.

Let us consider in the parameter space various sections of the stability region by planes $x=$ const.

Figure 2.41 shows in q, η plane, the stability regions limited by N_- and N_φ are given for $n=1$ and $x=0$. To shade the stability region, we use the hatching rule [47], which determines the side N_φ corresponding to the transition of the roots $z=e^{i\varphi}$ inside the unit circle. This side is determined by inequality

$$1>\frac{A_2\left(A_0^2-A_4^2\right)(A_0-A_4)-(A_0A_1-A_3A_4)(A_0A_3-A_1A_4)}{(A_2^2-A_4^2)^2-(A_0A_3-A_1A_4)^2}. \tag{2.5.42}$$

Substituting A_i coefficients into inequality (2.5.42), we obtain the hatching rule in the form

$$\eta>0,\ q>\frac{\eta^2\left|tg\frac{T}{2}\eta\right|}{|1-\eta^2|\left|\eta\frac{T}{2}-tg\frac{T}{2}\eta\right|}. \tag{2.5.43}$$

Obviously, a similar result could be obtained if in each region limited by N_- and N_φ we select some points and verify in them the Routh–Hurwitz stability conditions.

With increasing x the boundary N_φ in this plane is not changed, as it is a cylindrical surface along x. Equation (2.5.36) for N_- (2.5.36) represents a cylindrical surface with a generatrix parallel to axis q; therefore, in the plane q, η this part of the

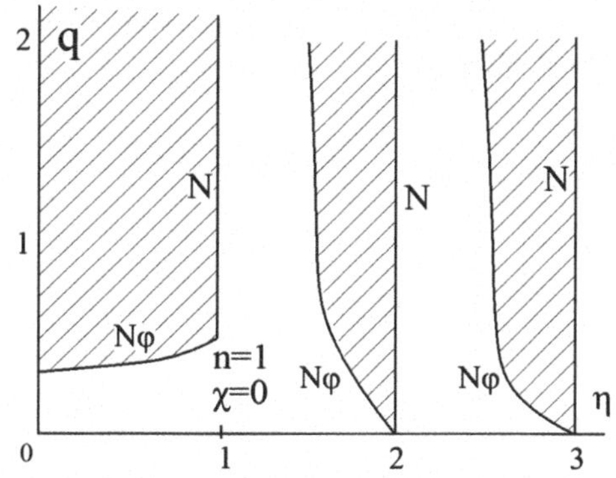

Fig. 2.41 Areas of stability of periodic motion modes

boundary N_- represents a series of lines $\eta = const$. These lines shift with increasing x toward smaller η, causing the narrowing of the corresponding stability regions. When $x \to \infty$, those lines tend to straight lines, the equations of which have the from

$$n\eta = \frac{1}{2} + \ell, \ell = 0, 1, 2, \ldots \tag{2.5.44}$$

For small x (which corresponds to a small impact mass and a large vibrator mass), the Eq. (2.5.37) for the boundary N_- does not yet give a boundary in the plane q, η, but with further increase in x along with the straight lines (2.5.36), the intersection of the x = const plane with the stability region occurs limited by the surface (2.5.37). With increase in x, this part of the stability region expands and descends into the region of smaller q. Figure 2.42 presents the stability regions limited by N_- and N_φ curves for large values x ($x = 50$) and $n = 1$. It is seen that the widest part of stability region lies for $\eta < 1$.

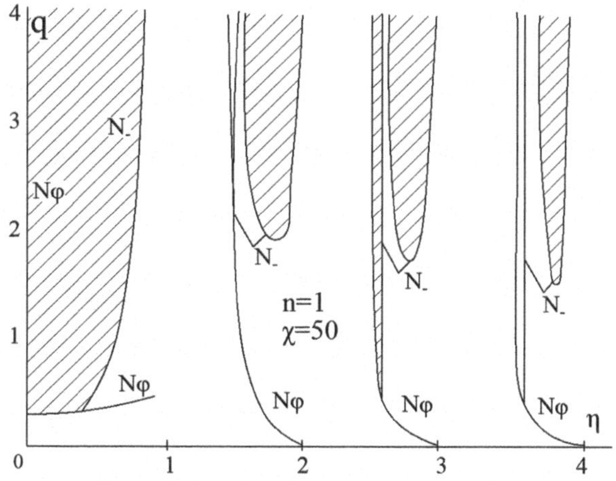

Fig. 2.42 Areas of stability of periodic motion modes

Thus, to find the most optimal parameter values for the system, a resonance curve is to be plotted

$$\lambda = \frac{1}{\sqrt{1+x}}$$

and parameters satisfying the inequality are to be chosen

$$\lambda^2(1+x) < 1. \tag{2.5.45}$$

The inequality (2.5.45) shows that to tune the system to optimal mode it is necessary to choose striker mass m_1, spring rigidity k and frequency of the driving force ω within

$$0 < \frac{k}{m_1\omega^2} < 1. \tag{2.5.46}$$

Hence, if the partial frequency of the system is less than unity, i.e., $\sqrt{\frac{k}{m_1}} < 1$, then the frequency of the external force ω can be chosen either greater than unity or less than unity, but when $\sqrt{\frac{k}{m_1}} > 1$, then ω should be greater than unity.

Within creasing n stability regions are getting much narrower, therewith, new stability regions are observed to appear. Thus, Fig. 2.43 shows the stability regions for $x = 0$ and $n = 2$, and it can be seen that as compared with stability regions for $n = 1$ (see Fig. 2.41) they are much narrower. Note that between the straight lines defined by the equations

$$\pi n x \eta ctg(\pi n \eta) = -1$$

with increasing n new straight lines and along with them new stability regions appear. When $n \to \infty$, the number of these straight lines tends to infinity.

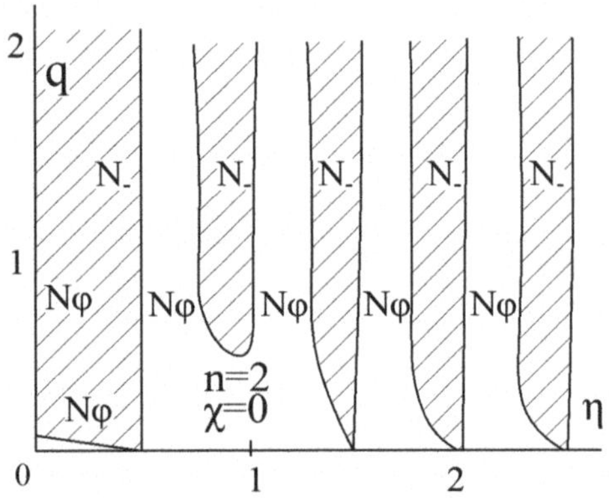

Fig. 2.43 Areas of stability of periodic motion modes

With increasing x, qualitative behavior of stability regions coincides with the case $n = 1$.

The bifurcation boundary equation C_τ for this case has the form

$$\left[-x + ctg(\pi n\eta)\sin(\eta\beta) - \cos\eta\beta + \frac{x\beta}{\pi n}\right]$$
$$= \left[\frac{1-\cos\eta\beta}{\eta}ctg(\pi n\eta) - x\beta - \frac{\sin(\eta\beta)}{\eta} + \frac{x\beta^2}{\pi n}\right]$$
$$q = \frac{x\eta^2|\sin\beta|}{|1-\eta^2|\pi n\left[-x + ctg(\pi n\eta)\sin(\eta\beta) - \cos\eta\beta + \frac{x\beta}{\pi n}\right]} \quad (2.5.47)$$

By solving the Eq. (2.5.47), it was shown that the boundary C_τ lies outside the stability region.

The qualitative picture of dividing the parameter space into stability regions for $R = 1$ is also preserved for the case when the coefficient of restitution R does not differ much from unity.

2.6 Conclusion

1. The operation of vibro-impact mechanisms in the mode of one impact per two (three, etc.) rotations of a vibrator shaft is characterized by a narrower stability interval (over ω) in comparison with the mode of one impact per rotation. For these reasons, the parameters of the mechanisms under study are recommended to be calculated on the basis of the condition one impact per one rotation.
2. In order to tune vibro-impact mechanisms, their parameters must be selected in the vicinity of the resonance frequency of the corresponding linear two-mass system without a limiter.
3. With an increase in the coefficient of restitution R upon impact, the stability regions of single-impact n-multiple periodic motions shift toward lower frequencies, become more indented, and the indentation appears from the side of negative gaps. In view of this, with increasing R in order to maintain periodic motions of this type, the initial gap ξ_0 should be increased.
4. Impact velocities in the stability regions at near resonant frequencies corresponding to linear model are significantly larger in magnitude than impact velocities in other regions. The maximum values of impact velocities for gaps that are not zero lie near the stability boundary (at negative gaps on the right boundary and at positive gaps on the left).
5. The developed methodology for studying the regions of existence and stability allows one to recommend in advance such values of the main parameters of the two-mass vibro-impact mechanism (the value of the lower and upper masses, the

rigidity of upper and lower springs, the frequency amplitude of the driving force) which ensure its stable operation in a single-impact periodic mode.
6. Software product makes it possible to speed up the process of processing experimental data, transfer research from field conditions to laboratory, and significantly reduce the complexity and costs associated with the experiment.
7. It is advisable to use the results obtained when creating a system of automatic tuning to the optimal operating mode, characterized by maximum impact velocity.

References

1. Savinov, O.A., Luskin, A.Y.: Vibration method of piling and its application in construction. Gosizdat, L-M (1960)
2. Golovachev, A.S., (ed.).: Studies of vibration and vibro-impact driving of piles. Transport, M. (1968)
3. Bespalova, L.V.: To the theory of vibro-impact mechanism. Izvestia of the Academy of Sciences of the USSR, OTN, No. 5 (1957)
4. Brogliato, B., Rio, A.Z.: On the control of complementary-slackness juggling mechanical systems. IEEE Trans. Autom. Control **45**(2), 235–246 (2000)
5. Ibrahim, R.A.: Vibro-impact dynamics. Modeling, Mapping and Applications, vol. 43, p. 297. Springer, Berlin Heidelberg (2009). https://doi.org/10.1007/978-3-642-00275-5
6. Bartolo, D., Boudaoud, A., Narcy, G., Bonn, D.: Dynamics of non-Newtonian droplets. Phys. Rev. Lett. **99**, 174502 (2007)
7. Brogliato, B.: Some perspectives on the analysis and control of complementarity systems. IEEE Trans. Autom. Control **48**(6), 918–935 (2003)
8. Brogliato, B., Lozano, R., Egeland, O.: Dissipative systems analysis and control. Springer (2007)
9. Babitsky, V.I., Krupenin, V.L.: Oscillations in Strongly Non-linear Systems, p. 320. Nauka, M. (1985)
10. Kurosh, A.G.: High Algebra. M. (1963)
11. Neimark, Y.I.: The method of point transformations in the theory of nonlinear oscillations. Izv. High. uch. Head, Radiophys. (1) (1958)
12. Neimark, Y.I.: The method of point mappings in the theory of nonlinear oscillations. Izv.vysh.uch. Head. Radiophys. **2**(2) (1958)
13. Neimark, Y.I.: Method of point transformations in the theory of nonlinear oscillations. Izv. High. uch. Head, Radiophys. **3**(5–6) (1958)
14. Bespalova, L.V., Neimark, Y.I., Feigin, M.I.: Dynamical systems with shock interactions and the theory of nonlinear oscillations. Eng. MTT Mag. (1) (1966)
15. Andronov, A.A., Witt, A.A., Khaikin, S.E.: Theory of Oscillations. Fizmatgiz, M. (1959)
16. Barkan, D.D., Shekhter, O.Y.: Forced vibrations of a vibrator with a movable limiter. JTF **25**(13) (1955)
17. Bykhovsky, I.I.: Fundamentals of the Theory of Vibration Technology. M. (1969)
18. Zaretsky, L.B.: Forced oscillations of a 6-spring vibro-impact mechanism with two impacts. Sat. Vibr. Tech. M. (1966)
19. Zaretsky, J.I.B.: On self-synchronization of centrifugal vibration exciters of vibroimpact mechanism. Mach. Sci. (1) (1967)
20. Peterka, F.: Periodicke kmitani Mechanicke seustarvy s vnitrnimi RAZY. Strojncky Cas. **16**(6) (1965)
21. Peterka, F.: Modelovani Kmitajich Soustav s razy on analogovam pocctaci. Strojnicky Cas. **16**(3) (1965)

22. Lukomsky, S.I.: Spring vibro-hammers and their main parameters. NIIM Stroydorkommunal, M. (1966)
23. Goldin, Y.M.: On the choice of the optimal scheme of the vibro-impact mechanism. Mach. Sci. (3) (1968)
24. Goldin, Y.M.: Comparative study of one-mass and two-mass vibro-impact mechanisms. On Sat. Research of Construction and Road Machines with the Use of Computer Technology. TsNIITE stroymash (1968)
25. Barkan, D.D., Shekhter, O.Y.: Vibro-impact mechanisms with a special impact mass. Proceedings of the Research Institute of Foundations and Foundations, Collection, No. 28, Gosstroy Publishing House (1956)
26. Kobrinsky, A.A., Kobrinsky, A.E.: Two-dimensional vibro-impact systems. Nauka, p. 336 (1981)
27. Kobrinsiy, A.E., Tyves, P.I.: Dynamics and stability of systems containing two shock pairs. Mach. Sci. (4) (1965)
28. Osmakov, S.A., Savinov, O.A.: Forced oscillations of a system of two masses interacting with the force of dry friction. Izv. Univ. Eng. (4), 126–134 (1963)
29. Ragulkis, K.M., Vitkus, I.I., Ragulskene, V.L.: Self-synchronization of mechanical systems. Self-synchronization and Vibro-Impact Systems. From Mintis, Vilnius (1965)
30. Batalova, Z.S.: On one nonlinear problem related to the theory of vibrational immersion. Izv.vysh.uch.zav. Radiophys. **6**(1) (1963)
31. Shlyakhtin, A.V.: Periodic motions of a nonlinear autonomous system / with an internal combustion engine. Tr. IMASH AS USSR **22**(88) (1961)
32. Boyarinov, V.S., Neimark, Y.I.: On shaft vibrations in a ball bearing. Izv. USSR Academy of Sciences, Mechanics, No. 3 (1965)
33. Bespalova, L.V.: On the theory of vibroimpact driving. Izv.vys. Academician, Radiophys. **2**(4) (1959)
34. Sijin, Z., Libiao, Z., Qishao, L.: A map method for grazing bifurcation in linear vibro–impact system. Acta Mech. Sinica **39**(1), 132–136 (2007) (in Chinese)
35. Bebikhov, V.A.: Modeling on MN-8 Vibrohammer. Izv. High. uch. Head
36. Babitsky, V.I., Izrailovich, M.Y.: On the optimal motions of vibroimpact systems. Mach. Sci. (3) (1968)
37. Althausen, E.E., Godes, Y.Yu.: Electronic impact model. Mach. Sci. (3) (1966)
38. Oldenburg, R., Sartorius, G.: Dynamics of Automatic Control. GEI M-L (1949)
39. Karpov, N.A., Lobanov, V.I., Tsitel'nyi, P.M.: Self-synchronizing percussive vibration machine. Proceedings of the Interuniversity Scientific Conference on Electric Percussion Machines, Novosibirsk (1967)
40. Althausen, E.V., Godes, Y.Y.: Modeling shock systems. Mach. Sci. (3) (1968)
41. Ragulskene, V.L., Satkevechus, E.E.: Electronic modeling of multimass vibro-impact systems. Tr. Higher uch. Head Lit.SSR, Vibro-Technics, No. 1 (1967)
42. Bespalova, L.V., Lobanov, V.I., Metrikin, V.S.: The results of the calculation on the computer of the boundaries of the stability of the movement of the vibro-impact compactor of the slope of the prism. Bulletin of the All-Union Research Institute of Railway. dor. Transport, No. 1 (1969)
43. Bykhovsky, I.I.: On the parameters of vibratory hammers and vibratory hammers. On Sat. Pile Equipment, M., TsINTIAM (1964)
44. Hayashi, T.: Forced Oscillations in Nonlinear Systems, Ed. I.L., M. (1957)
45. Bykhovsky, I.I., Dorohova, A.D., Zaretsky, L.B., Lukomsky, S.I.: On some periodic motions and the structure of the phase space of a shock-oscillatory system with a constant restoring force. Izv. USSR Academy of Sciences, OTV, Mechanics and Engineering, No. 2 (1964)
46. Feigin, M.I.: On the construction of the stability region in the space of parameters of a dynamical system. PMM **32**(1) (1968)
47. Brusin, V.A., Tai, M.J.I.: Absolute stochastic stability. Izv. High. uch. Head, Radiophys. **10**(7) (1967)

Chapter 3
Research into Multi-mass Vibro-impact System

Abstract In the second chapter, a study is carried out of a multi-mass vibration-impact system, which is a generalization of a two-mass system and consists of $N > 2$ identical masses connected by springs of the same stiffness. Under these assumptions, it was possible, by introducing normal coordinates, to decompose the system into N second-order equations and, in a compact form, to find in analytical form the coordinates of fixed points corresponding to single-impact n-multiple periodic movements and to study their stability depending on the stiffness values N of the connecting springs, the amplitude and frequency of the disturbing force. The problem of studying the stability of fixed points is reduced to determining the roots of some characteristic equation that satisfy the conditions $|z_i| < 1$. This problem is reduced to finding the maximum absolute value of the matrix A eigenvalue belonging to the interior of a circle of unit radius. The optimal values of the system parameters are found at which the maximum modulus eigenvalue of the matrix A (or the maximum modulus root of the characteristic polynomial) takes on a minimum value.

Keywords Mathematical model · Multi-mass vibration-impact mechanism · Matrix eigenvalues · Stability · Fixed point · Point mapping method · Normal coordinates

3.1 Equation of Motion. Finding the Invariant Points of Point Mapping

Studying the properties of a multi-mass vibro-impact system consisting of N masses m_i interconnected by means of springs with stiffness coefficients C_i is of theoretical and practical interest. The results obtained for sufficiently large N can be used as first approximations and for the study of distributed systems, e.g., piles, beams, etc.

Vibro-impact systems with many $N > 2$ impact elements were studied in [1–6] and other works, from which it directly follows that dynamics only in some [1] specific systems can be investigated without using computer simulation.

V. Metrikin and I. Leonid, *Forced Oscillations of Multidimensional Highly Nonlinear Systems*, Advanced Structured Materials 222,
https://doi.org/10.1007/978-981-97-6779-3_3

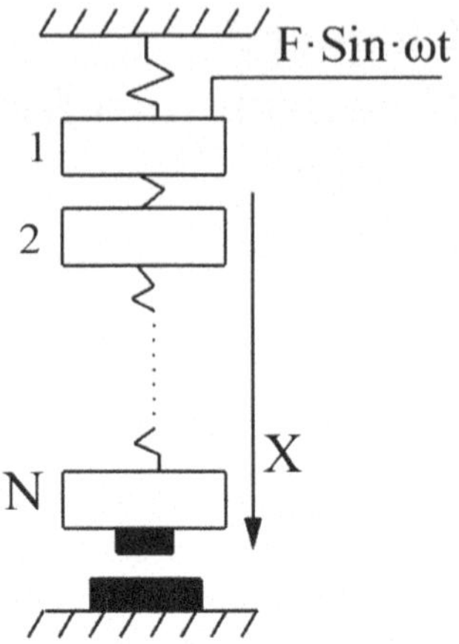

Fig. 3.1 Physical diagram of a multi-mass vibration-impact mechanism

We study the dynamics of a vibro-impact system, consisting of N identical masses $m_i = m,\ i = 1, 2, \ldots N$ interconnected by springs with a stiffness coefficient $C_i = C,\ i = 1, 2, \ldots, N$ (Fig. 3.1). An external periodic force $F \sin \omega t$ acts on the mass m_1 which is elastically connected with a fixed support. Assuming that the springs operate in the linear part of their characteristics, the equations of motion of the system in the intervals between impacts have the following form (the position of the static equilibrium of the system is chosen for the reference point)

$$\begin{aligned} &m\ddot{x}_1 + c(2x_1 - x_2) = mg + F \sin \omega t \\ &m\ddot{x}_N + c(2x_N - x_{N-1}) = mg \\ &m\ddot{x}_n + c(2x_n - x_{n+1} - x_{n-1}) = mg \end{aligned} \tag{3.1.1}$$

Note that these equations can be obtained from the general partial differential equation for the vibration of a distributed bar. The equations of oscillation of the bar have the form

$$\frac{\partial^2 u}{\partial t^2} - c^2 \frac{\partial^2 u}{\partial x^2} = 0 \tag{3.1.2}$$

Applying approximate formulas for the second derivatives in the form

$$\frac{\partial^2 u}{\partial x^2} = \frac{1}{h}(u_{n+1} + u_{n-1} - 2u_n) \tag{3.1.3}$$

we directly obtain the average equation of the system (3.1.1). Using the boundary conditions, we can obtain the two remaining equations.

Introducing dimensionless variables in the system

$$\xi_i = \frac{m\omega^2}{F}\left(x_i - x_i^0\right), \ \omega t = \tau \tag{3.1.4}$$

it is rewritten in the form

$$\begin{aligned}
&\ddot{\xi}_1 + \lambda^2(2\xi_1 - \xi_2) = \sin\tau \\
&\ddot{\xi}_n + \lambda^2(2\xi_n - \xi_{n+1} - \xi_{n-1}) = 0, \quad n = 2, 3, \ldots, N-1 \\
&\ddot{\xi}_N + \lambda^2(\xi_N - \xi_{N-1}) = 0
\end{aligned} \tag{3.1.5}$$

where $\lambda^2 = C/m\omega^2$ and values x_i^0 determine the static equilibrium position of the system, and are found as a solution to the system of linear equations

$$\begin{aligned}
&2x_i^0 - x_{i+1}^0 - x_{i-1}^0 = \ell_0 \\
&2x_1^0 - x_2^0 = \ell_0 \\
&x_N^0 - x_{N-1}^0 = \ell_0
\end{aligned} \tag{3.1.6}$$

where ℓ_0 denotes mg/c.

The general solution of the homogeneous system (3.1.5) will be sought in the form of the following combination

$$\xi_n = \sin\alpha n e^{-i\Lambda t} \tag{3.1.7}$$

Substituting (3.1.7) into the second equation of the system (3.1.5), we obtain

$$0 = -\Lambda^2 \sin\alpha n + \lambda^2[2\sin\alpha n - \sin\alpha(n+1) - \sin\alpha(n-1)] \tag{3.1.8}$$

After some transformation (3.1.8), we derive

$$4\lambda^2 \sin^2\frac{\alpha}{2} = \Lambda^2 \tag{3.1.9}$$

Substituting (3.1.7) into the third Eq. (3.1.5), we obtain

$$-\Lambda^2 \sin\alpha N + \lambda^2[\sin\alpha n - \sin\alpha(N-1)] = 0$$

Using trigonometric formulas, we will have

$$-\Lambda^2 \sin\alpha N + 2\lambda^2 \sin\alpha N \sin^2\frac{\alpha}{2} + \lambda^2 \cos\alpha N \sin\alpha = 0$$

Using further the relation (3.1.9), we obtain the equation

$$\cos\alpha(N + 1/2) = 0 \tag{3.1.10}$$

The solution of which can be written in the form

$$\alpha_k = \frac{\pi(k - 1/2)}{N + 1/2} \quad (k = 1, 2, \ldots, N) \tag{3.1.11}$$

Thus, the general solution of the homogeneous system (3.1.5) is written in the form

$$\xi_n = \sum_{k=1}^{N} A_k \ell^{-i\Lambda\tau} \sin\alpha_k n = \sum_{k=1}^{N} A_k(t) \sin\alpha_k n \tag{3.1.12}$$

where α_k is derived from (3.1.11) and A_k is normal coordinates

$$A_k = \frac{1}{a} \sum_{n=1}^{N} \xi_n \sin\alpha_k n \tag{3.1.13}$$

Then using the known relations [7]

$$\frac{1}{a^2} \sum_{k=1}^{N} \sin\alpha_k n \sin\alpha_k n' = \delta_{nn'}, \left(a^2 = \frac{2N+1}{4}\right)$$

$$\frac{1}{a^2} \sum_{k=1}^{N} \sin\alpha_k n \sin\alpha_{k'} n = \delta_{kk'} \tag{3.1.14}$$

and introducing the matrix

$$\|\beta_{ik}\| = \lambda^2 \begin{Vmatrix} 2 & -1 & 000 & \ldots\ 00 \\ -1 & 2 & 100 & \ldots\ 00 \\ 0 & -1 & 2-10 & \ldots\ 00 \\ \ldots & \ldots & \ldots & \ldots\ldots \\ 0 & 0 & 000 & \ldots -1\ +1 \end{Vmatrix} \tag{3.1.15}$$

The system of equtions (3.1.5) will be rewritten in the form

$$\ddot{\xi}_k + \beta_{ik}\delta_{kk'} = \delta_{i1} \sin\tau, \quad i = 1, 2, \ldots, N \tag{3.1.16}$$

Using (3.1.13), the system (3.1.16) is rewritten in terms of normal coordinates

$$\ddot{A}_e + \Lambda_e^2 A_e = \sin\tau \sin\alpha_e, \quad e = 1, 2, \ldots, N \tag{3.1.17}$$

The system of equations (3.1.16) in terms of normal coordinates has been transformed into N independent equations for variable A.

Equation (3.1.17) with initial conditions at $\tau = \tau_0, A_n = A_n^0, \dot{A}_n = \dot{A}_n^0, A_N = \xi_0, \dot{A}_N = \dot{A}_N^+$ can be written

$$\begin{aligned} A_n &= \Lambda_n^{-1} \dot{A}_n^+ \sin\Lambda_n(\tau - \tau_0) + A_n^0 \cos\Lambda_n(\tau - \tau_0) \\ &+ \zeta_n a^{-1} \sin\alpha_n[in\tau - \sin\tau_0 \cos\Lambda_n(\tau - \tau_0) \end{aligned}$$

$$
\begin{aligned}
&-\Lambda_n^{-1}\cos\tau_0\sin\Lambda_n(\tau-\tau_0)\big]\\
\dot{A}_n &= \dot{A}_n^{+}\cos\Lambda_n(\tau-\tau_0) - A_n^0\Lambda_n\sin\Lambda_n(\tau-\tau_0)\\
&+\zeta_n a^{-1}\sin\alpha_n[\cos\tau+\Lambda_n\sin\tau_0\sin\Lambda_n(\tau-\tau_0)\\
&-\cos\tau_0\cos\Lambda_n(\tau-\tau_0)]
\end{aligned}
\tag{3.1.18}
$$

where $\zeta_n = (\Lambda_n^2 - 1)^{-1}$.

The solution (3.1.18) is valid in the intervals between impacts. For a complete description of the motion of the system, it is necessary to add the impact conditions at time $\tau = \tau_0$.

$$
\begin{aligned}
&\text{at } \xi_N(\tau_0) = \xi_0\\
&\dot{\xi}_N(\tau_0+0) = -R\dot{\xi}_N(\tau_0-0)
\end{aligned}
\tag{3.1.19}
$$

The phase space of the system is $(2N+1)$—dimensional in coordinates $\xi_i, \dot{\xi}_i, \tau (i = 1, 2, \ldots, N)$. The plane of the limiter in the phase space corresponds to the surface Π, which is defined by the equation $\xi_N = \xi_0$. The whole phase space is partitioned by the surface Π into two subspaces $G(\xi_N \le \xi_0)$ and $W(\xi_N > \xi_0)$. All possible motions of the system are in the subspace G. In the case of impact motions, the research into the system behavior is reduced to studying the point mapping S of the surface Π into itself.

In the space of normal coordinates $A_k, \dot{A}_k, \tau$, the surface Π is determined by the equation

$$
\sum_{k=1}^{N} A_k \sin\alpha_k N = a\xi_0 \tag{3.1.20}
$$

Thus, point mapping S is derived from (3.1.18) with additional conditions (3.1.20) in the form

$$
\begin{aligned}
A_n^1 &= \Lambda_n^{-1}\dot{A}_n^{+}\sin\Lambda_n(\tau_1-\tau_0) + A_n^0\cos\Lambda_n(\tau_1-\tau_0)\\
&+\zeta_n a^{-1}\sin\alpha_n[\sin\tau_1 - \sin\tau_0\cos\Lambda_n(\tau_1-\tau_0)\\
&-\Lambda_n^{-1}\cos\tau_0\sin\Lambda_n(\tau-\tau_0)\big]\\
\dot{A}_n^1 &= \dot{A}_n^{+}\cos\Lambda_n(\tau_1-\tau_0) - A_n^0\Lambda_n\sin\Lambda_n(\tau_1-\tau_0)\\
&+\zeta_n a^{-1}\sin\alpha_n[\cos\tau_1+\Lambda_n\sin\tau_0\sin\Lambda_n(\tau_1-\tau_0)\\
&-\cos\tau_0\cos\Lambda_n(\tau_1-\tau_0)]
\end{aligned}
$$

$$
\sum_{k=1}^{N} A_k \sin\alpha_k N = a\xi_0 \tag{3.1.21}
$$

A periodic motion in the phase space corresponds to a fixed point of the map S (points differing in τ by a multiple 2π are identified). The coordinates of the fixed point of a single-impact n-fold periodic mode are determined from the S map (3.1.21) with the addition of periodicity conditions $A_k^1 = A_k^0, \dot{A}_k^1 = \dot{A}_k^0, \tau_1 - \tau_0 = 2\pi r = T$, namely

$$\begin{aligned} A_n^0 &= \Lambda_n^{-1}\dot{A}_n^+ \sin T\Lambda_n + A_n^0 \cos T\Lambda_n + \zeta_n a^{-1} \sin\alpha_n \\ &\quad \times \left[\sin\tau_0(1-\cos T\Lambda_n) - \Lambda_n^{-1}\cos\tau_0 \sin T\Lambda_n\right] \\ \dot{A}_n^0 &= \dot{A}_n^+ \cos T\Lambda_n - A_n^0\Lambda_n \sin T\Lambda_n + \zeta_n a^{-1} \sin\alpha_n \\ &\quad \times \left[\Lambda_n \sin\tau_0 \sin T\Lambda_n + \cos\tau_0(1-\cos T\Lambda_n)\right] \end{aligned} \tag{3.1.22}$$

Substituting $\dot{A}_n^+$ from the first equation into the second, we obtain

$$\dot{A}_n^0 = -\dot{A}_n^+ + 2\zeta_n a^{-1} \sin\alpha_n \cos\tau_0 \tag{3.1.23}$$

Substituting $\dot{A}_n^+$ from (3.1.23) into impact condition

$$\dot{\xi}_n^+ = \begin{cases} \dot{\xi}_n^0 \\ -R\dot{\xi}_N^0 \end{cases}, \quad n = n = 1, 2, \ldots, N-1 \tag{3.1.24}$$

we derive

$$\begin{aligned} a^{-1}\sum_{n=1}^{N} \dot{A}_n^0 \sin\alpha_n k &= \sum_{n=1}^{N} \zeta_n \frac{\sin\alpha_n \sin\alpha_n k}{a^2} \cos\tau_0 \\ &\quad + \begin{cases} 0, \ k = 1, 2, \ldots, N-1 \\ \frac{1+R}{1-R} \sum\limits_{n=1}^{N} \zeta_n a^{-2} \sin\alpha_n N \cos\tau_0 \end{cases} \end{aligned} \tag{3.1.25}$$

Equation (3.1.25) is obtained in a non-standard way; therefore, we briefly explain how it was derived.

The impact condition (3.1.24) in normal coordinates will be rewritten in the form

$$\frac{1}{a}\sum_{n=1}^{N} \dot{A}_n^+ \sin\alpha_n k = \begin{cases} \frac{1}{a}\sum\limits_{n=1}^{N} \dot{A}_n^0 \sin\alpha_n k, \quad k = 1, 2, \ldots, N-1 \\ -R\frac{1}{a}\sum\limits_{n=1}^{N} \dot{A}_n^0 \sin\alpha_n N, \quad k = N \end{cases} \tag{3.1.26}$$

Substituting (3.1.23) into (3.1.26), we obtain

$$\frac{1}{a}\sum_{n=1}^{N}\left[-\dot{A}_n^0 + 2\zeta_n a^{-1} \sin\alpha_n \cos\tau_0\right] \sin\alpha_n k$$

$$= \begin{cases} a^{-1} \sum\limits_{n=1}^{N} \dot{A}_n^0 \sin \alpha_n k, & k = 1, 2, \ldots, N-1 \\ -Ra^{-1} \sum\limits_{n=1}^{N} \dot{A}_n^0 \sin \alpha_n N, & k = N \end{cases} \tag{3.1.27}$$

Equation (3.1.27) can be written in the form

$$-a^{-1} \sum_{n=1}^{N} \dot{A}_n^0 \sin \alpha_n k = -2 \sum_{n=1}^{N} \zeta_n a^{-2} \sin \alpha_n \sin \alpha_n k \cos \tau_0 + \begin{cases} a^{-1} \sum\limits_{n=1}^{N} \dot{A}_n^0 \sin \alpha_n k \\ -Ra^{-1} \sum\limits_{n=1}^{N} \dot{A}_n^0 \sin \alpha_n N \end{cases} \tag{3.1.28}$$

Transposing $a^{-1} \sum_{n=1}^{N} \dot{A}_n^0 \sin \alpha_n N$ from the right-hand to the left-hand side, we obtain

$$-2a^{-1} \sum_{n=1}^{N} \dot{A}_n^0 \sin \alpha_n k = -2 \sum_{n=1}^{N} \zeta_n a^{-2} \sin \alpha_n k \cos \tau_0 + \begin{cases} 0, & k = 1, 2, \ldots, N-1 \\ -(1+R)a^{-1} \sum\limits_{n=1}^{N} \dot{A}_n^0 \sin \alpha_n N, & k = N \end{cases} \tag{3.1.29}$$

From relations (3.1.29), it can be derived that for $k = 1, 2, \ldots, N-1$, the following equation is satisfied

$$\sum_{n=1}^{N} \dot{A}_n^0 \sin \alpha_n k = \sum_{n=1}^{N} \zeta_n a^{-1} \sin \alpha_n \sin \alpha_n k \cos \tau_0 \tag{3.1.30}$$

and for $k = N$,

$$\sum_{n=1}^{N} \dot{A}_n^0 \sin \alpha_n N = \frac{2}{1-R} \sum_{n=1}^{N} \zeta_n a^{-1} \sin \alpha_n \sin \alpha_n N \cos \tau_0 \tag{3.1.31}$$

is satisfied.

Using (3.1.29)–(3.1.30), we derive (3.1.25). From (3.1.30) and (3.1.31), it can be written that

$$\begin{aligned} \dot{A}_n^0 &= \frac{\cos \tau_0}{a}[\zeta_n \sin \alpha_n + \frac{1+R}{1-R} g \sin \alpha_n N \\ \dot{A}_n^+ &= \frac{\cos \tau_0}{a}[\zeta_n \sin \alpha_n - \frac{1+R}{1-R} g \sin \alpha_n N \end{aligned} \tag{3.1.32}$$

where the following notation is introduced

$$g=\sum_{n=1}^{N} a^{-2}\zeta_n \sin\alpha_n \sin\alpha_n N \tag{3.1.33}$$

From (3.1.22), it can be derived that

$$A_n^0=a^{-1}\left(\zeta_n \sin\alpha_n \sin\tau_0-\cos\tau_0 g\frac{1+R}{1-R}\frac{\sin\alpha_n N}{\Lambda_n}Ag\frac{T}{2}\Lambda_n\right) \tag{3.1.34}$$

Using impact condition (3.1.20) and relation (3.1.33), (3.1.34), we derive that

$$g=\left[\sin\tau_0-\frac{1+R}{1-R}\ell\cos\tau_0\right]=\xi_0 \tag{3.1.35}$$

where ℓ denotes

$$\ell=\sum_{n=1}^{N} a^{-2}\Lambda_n^{-1}\sin^2\alpha_n Nctg\frac{T}{2}\Lambda_n \tag{3.1.36}$$

From Eq. (3.1.35), we obtain the condition of the existence of an in variant point S transformation in the form

$$|\xi_0|=|g|\sqrt{1+[(1+R)/(1-R)]^2} \tag{3.1.37}$$

Going from normal coordinates to variables ξ, $\dot{\xi}$, we obtain the coordinates of a fixed point in the following form

$$\begin{aligned}
\xi_i^0 &= A_i^6\sin\tau_0-\frac{1+R}{1-R}A_N^6\ell_i^1\cos\tau_0\\
\dot{\xi}_i^0 &= A_i^6\cos\tau_0+\frac{1+R}{1-R}A_N^6\delta_{iN}\cos\tau_0\\
\xi_0 &= A_N^6\left(\sin\tau_0-\frac{1+R}{1-R}\ell_N^1\cos\tau_0\right)
\end{aligned} \tag{3.1.38}$$

where

$$\begin{aligned}
\ell_i^1 &= a^{-2}\sum_{n=1}^{N}\sin\alpha_n i\sin\alpha_n N\frac{ctg\frac{T}{2}\Lambda_n}{\Lambda_n}\\
A_i^6 &= a^{-2}\sum_{n=1}^{N}\zeta_n\sin\alpha_n i\sin\alpha_n
\end{aligned} \tag{3.1.39}$$

From the problem statement, it follows that the impact velocity is always positive, therefore, we obtain another boundary of positivity

$$\frac{2}{1-R}A_N^6 \cos \tau_0 \geq 0 \tag{3.1.40}$$

3.2 Stability of Invariant Points of a Point Mapping. Finding Optimal Parameter Values

To study the stability of a fixed point of a point mapping S, we linearize the equations of a point mapping in the vicinity of a fixed point and then assuming $\delta\{M_1\} = z\delta\{M_0\}$, S-mapping in matrix form can be written in the form

$$\begin{pmatrix} A_n^1 - \beta_n \sin \tau_1 \\ \dot{A}_n^1 - \beta_n \cos \tau_1 \end{pmatrix} = \begin{pmatrix} \cos \Lambda_n(\tau_1 - \tau_0) & \Lambda_n^{-1} \sin \Lambda_n(\tau_1 - \tau_0) \\ -\Lambda_n \sin \Lambda_n(\tau_1 - \tau_0) & \cos \Lambda_n(\tau_1 - \tau_0) \end{pmatrix} \times \begin{pmatrix} A_n^0 - \beta_n \sin \tau_0 \\ \dot{A}_n^+ - \beta_n \cos \tau_0 \end{pmatrix} \tag{3.2.1}$$

where β_n denotes $a^{-1}\zeta_n \sin \alpha_n$.

Let us vary Eqs. (3.2.1) in the neighborhood of a fixed point:

$$\begin{pmatrix} \delta A_n^1 - \beta_n \cos \tau_0 \delta\tau_1 \\ \delta\dot{A}_n^1 + \beta_n \sin \tau_0 \delta\tau_1 \end{pmatrix} = \begin{pmatrix} \cos T\Lambda_n & \Lambda_n^{-1} \sin T\Lambda_n \\ -\Lambda_n \sin T\Lambda_n & \cos T\Lambda_n \end{pmatrix} \begin{pmatrix} \delta A_n^0 - \beta_n \cos \tau_0 \delta\tau_0 \\ \delta\dot{A}_n^+ + \beta_n \sin \tau_0 \delta\tau_0 \end{pmatrix} + \begin{pmatrix} -\Lambda_n \sin T\Lambda_n & \cos T\Lambda_n \\ -\Lambda_n^2 \cos T\Lambda_n & -\Lambda_n \sin T\Lambda_n \end{pmatrix} \begin{pmatrix} A_n^0 - \beta_n \sin \tau_0 \\ \dot{A}_n^+ - \beta_n \cos \tau_0 \end{pmatrix} \delta(\tau_1 - \tau_0) \tag{3.2.2}$$

Using expressions for the coordinates of a fixed point, the second term in the right-hand side of the equation is rewritten in the form

$$\frac{1+R}{1-R} g \sin \alpha_n N \cos \tau_0 \begin{pmatrix} 1 \\ \Lambda_n ctg \frac{T}{2} \Lambda_n \end{pmatrix} (\delta(\tau_1) - \delta\tau_0) \tag{3.2.3}$$

Then by substituting (3.2.3) into (3.2.2) and grouping the terms with the same variations, we obtain

$$\begin{pmatrix} \delta A_n^1 \\ \delta\dot{A}_n^1 \end{pmatrix} = \beta_n \begin{pmatrix} \cos \tau_0 \\ -\sin \tau_0 \end{pmatrix} \delta\tau_1 + \frac{1+R}{1-R} g \cos \tau_0 \frac{\sin \alpha_n N}{a} \begin{pmatrix} 1 \\ \Lambda_n ctg \frac{T}{2} \Lambda_n \end{pmatrix} \delta(\tau_1 - \tau_0)$$

$$-\beta_n\begin{pmatrix}\cos T\Lambda_n & \Lambda_n^{-1}\sin T\Lambda_n\\ -\Lambda_n\sin T\Lambda_n & \cos T\Lambda_n\end{pmatrix}\begin{pmatrix}\cos\tau_0\\ -\sin\tau_0\end{pmatrix}\delta\tau_0$$

$$+\begin{pmatrix}\cos T\Lambda_n & \Lambda_n^{-1}\sin T\Lambda_n\\ -\Lambda_n\sin T\Lambda_n & \cos T\Lambda_n\end{pmatrix}\begin{pmatrix}\delta A_n^0\\ \delta\dot{A}_n^+\end{pmatrix} \tag{3.2.4}$$

For further research, we introduce three matrices of size $(n \times n)$ and five vectors of length B_i, ℓ_i in the following way

$$A_{ik}^1 = a^{-2}\sum_{n=1}^{N\sum_n \cos T_n} \sin\alpha_n i \sin$$

$$A_{ik}^2 = a^{-2}\sum_{n=1}^{N\sum_n \sin T_{nn}^{-1}} \sin\alpha_n i \sin$$

$$A_{ik}^3 = a^{-2}\sum_{n=1}^{N\sum_{nn} \sin T_n} \sin\alpha_n i \sin$$

$$B_i^1 = a^{-1}\sum_{n=1}^{N}\beta_n \sin\alpha_n i \cos T\Lambda_n$$

$$B_i^2 = a^{-1}\sum_{n=1}^{N}\beta_n \sin\alpha_n i\Lambda_n^{-1} \sin T\Lambda_n$$

$$B_i^3 = a^{-1}\sum_{n=1}^{N}\beta_n \sin\alpha_n i\Lambda_n \sin T\Lambda_n$$

$$B_i^4 = A_i^6$$

$$\ell_i^2 = a^{-2}\sum_{n=1}^{N\sum_{nn}\frac{T}{2}n} \sin\alpha_n i \sin \tag{3.2.5}$$

In (3.2.4), going from normal coordinates to the original dimensionless variables, we will have for $i = 1, 2, 3, \ldots N - 1$.

$$\begin{pmatrix}\delta\xi_i\\ \delta\dot{\xi}_i\end{pmatrix} = \begin{pmatrix}A_{ik}^1 & A_{ik}^2\\ -A_{ik}^3 & A_{ik}^1\end{pmatrix}\begin{pmatrix}\delta\xi_k^0\\ \delta\dot{\xi}_k^+\end{pmatrix} + \left[B_i^4\begin{pmatrix}\cos\tau_0\\ -\sin\tau_0\end{pmatrix} + \frac{1+R}{1-R}B_N^4\cos\tau_0\begin{pmatrix}\delta_{iN}\\ \ell_i^2\end{pmatrix}\right]\delta\tau_1$$

$$-\left[\begin{pmatrix}B_i^1 & B_i^2\\ -B_i^3 & B_i^1\end{pmatrix}\begin{pmatrix}\cos\tau_0\\ -\sin\tau_0\end{pmatrix} + \frac{1+R}{1-R}B_N^4\cos\tau_0\begin{pmatrix}\delta_{iN}\\ \ell_i^2\end{pmatrix}\right]\delta\tau_0 \tag{3.2.6}$$

and for $i = N$

$$\begin{pmatrix} 0 \\ \delta\dot{\xi}_N \end{pmatrix} = \begin{pmatrix} A^1_{Nk} & A^2_{Nk} \\ -A^3_{Nk} & A^1_{Nk} \end{pmatrix}\begin{pmatrix} \delta\xi^0_k \\ \delta\dot{\xi}^+_k \end{pmatrix} + \begin{pmatrix} 0 & A^2_{NN} \\ 0 & A^1_{NN} \end{pmatrix}\begin{pmatrix} 0 \\ -R\delta\dot{\xi}^0_N \end{pmatrix}$$
$$+ \begin{pmatrix} \frac{2}{1-R}\cos\tau_0 \\ -\sin\tau_0 + \frac{1+R}{1-R}\cos\tau_0\ell^2_N \end{pmatrix} B^4_i\delta\tau_1$$
$$- \left[\begin{pmatrix} B^1_N & B^2_N \\ -B^3_N & B^1_N \end{pmatrix}\begin{pmatrix} \cos\tau_0 \\ -\sin\tau_0 \end{pmatrix} + \begin{pmatrix} 1 \\ \ell^2_N \end{pmatrix}\frac{1+R}{1-R}B^4_N\cos\tau_0\right]\delta\tau_0 \quad (3.2.7)$$

By introducing further, a new variable y in the form

$$\delta y = -B^4_N\frac{2}{1-R}\cos\tau_0\delta\tau \quad (3.2.8)$$

from Eq. (3.2.7), we will have

$$\delta y = A^1_{Nk}\delta\xi^0_k + A^2_{Nk}\delta\dot{\xi}^0_k + \frac{1-R}{2}\left(\frac{B^1_N}{B^4_N} + \frac{1+R}{1-R} - \frac{B^2_N}{B^4_N}tg\tau_0\right)\delta y_0 - RA^2_{NN}\delta\dot{\xi}^0_N \quad (3.2.9)$$

Let us rewrite the matrix equation (3.2.6) for the new variable y in the following variations

$$\begin{pmatrix} \delta\xi_i \\ \delta\dot{\xi}_i \end{pmatrix} = \begin{pmatrix} A^1_{ik} & A^2_{ik} \\ -A^3_{ik} & A^1_{ik} \end{pmatrix}\begin{pmatrix} \delta\xi^0_k \\ \delta\dot{\xi}^+_k \end{pmatrix} + \begin{pmatrix} A^1_{iN} & A^2_{iN} \\ -A^3_{iN} & A^1_{iN} \end{pmatrix}\begin{pmatrix} 0 \\ -R\delta\dot{\xi}^0_N \end{pmatrix}$$
$$- \frac{1-R}{2}\left[\frac{B^4_i}{B^4_N}\begin{pmatrix} 1 \\ -tg\tau_0 \end{pmatrix} + \frac{1+R}{1-R}\begin{pmatrix} 0 \\ \ell^2_i \end{pmatrix}\right]\delta y$$
$$+ \frac{1-R}{2}\left[\frac{1}{B^4_N}\begin{pmatrix} B^1_i & B^2_i \\ -B^3_i & B^1_i \end{pmatrix}\begin{pmatrix} 1 \\ -tg\tau_0 \end{pmatrix} + \frac{1+R}{1-R}\begin{pmatrix} 0 \\ \ell^2_i \end{pmatrix}\right]\delta y_0 \quad (3.2.10)$$

Expanding Eq. (3.2.10), we obtain

$$\begin{aligned} \delta y &= A^1_{Nk}\delta\xi^0_k + A^2_{Nk}\delta\dot{\xi}^0_k - RA^2_{NN}\delta\dot{\xi}^0_N \\ &\quad + \Gamma^3_N\delta y_0, \quad k = 1, 2, \ldots, N-1 \\ \delta\xi_i &= (A^1_{ik} - \Gamma^1_i A^1_{Nk})\delta\xi^0_k + (A^3_{ik} - \Gamma^1_i A^2_{Nk})\delta\dot{\xi}^0_k \\ &\quad - R(A^2_{iN} - \Gamma^1_i A^2_{NN})\delta\dot{\xi}^0_N \\ &\quad + \left[\Gamma^3_i - \Gamma^1_i\Gamma^3_N - \frac{1+R}{2}\right]\delta y_0, \quad k, i = 1, 2, \ldots, N-1 \\ \delta\dot{\xi}_i &= (-A^3_{ik} - \Gamma^2_i A^1_{Nk})\delta\xi^0_k + (A^1_{ik} - \Gamma^2_i A^2_{Nk})\delta\dot{\xi}^0_k \end{aligned}$$

$$
\begin{aligned}
&- R\big(A_{iN}^{1} - \Gamma_i^2 A_{NN}^{2}\big)\delta\dot{\xi}_N^0 + \big[\Gamma_i^4 - \Gamma_i^2\Gamma_N^3\big]\delta y_0 \\
&k = 1, 2, \ldots, N-1 \\
&i = 1, 2, \ldots, N
\end{aligned}
\tag{3.2.11}
$$

where Γ_i denotes

$$
\begin{aligned}
\Gamma_i^1 &= \frac{1-R}{2}\frac{B_i^4}{B_N^4},\ \Gamma_i^2 = \frac{1}{2}\left[(1+R)\ell_i^2 - (1-R)\frac{B_i^4}{B_N^4}tg\,\tau_0\right], \\
\Gamma_i^3 &= \frac{1}{2}\left[(1+R) + (1-R)\frac{B_i^1 - B_i^2 tg\,\tau_0}{B_N^4}\right], \\
\Gamma_i^4 &= \frac{1}{2}\left[(1+R)\ell_i^2 - (1-R)\frac{B_i^3 + B_i^1 tg\,\tau_0}{B_N^4}\right]
\end{aligned}
\tag{3.2.12}
$$

Thus, the question of stability of a fixed point has been reduced to the determination of eigenvalues z_i in the eigenvalue problem (3.2.11), written in the general form

$$
Ax = zx \tag{3.2.13}
$$

where A is matrices in variations (3.2.11) and x is variation column.

A fixed point of the S-mapping will be stable if all the eigenvalues of the matrix A satisfy the condition [8–10]

$$
|z_i| < 1, \quad i = 1, 2, \ldots \tag{3.2.14}
$$

To determine the stability of the fixed point, it is not necessary to find all the eigenvalues of the matrix A, but to determine the largest modulo value, then (3.2.14) can be rewritten in the form of the inequality

$$
z^* = \max_i |z_i| < 1
$$

Using the above formulas, the parameter space was partitioned into stability regions for various N with the help of computer modeling.

To find the maximum modulus of the eigenvalue z^*, the power iteration was used [11]

Figures 3.1, 3.2 and 3.3 show the stability regions in $1/\lambda$, ξ_0 plane for the system consisting of one, two, and four masses, respectively (i.e., $N = 1, 2$ and 4). The coefficient of restitution at impact was 0.6. To optimally tune the system, it is necessary to take into account both the maximum impact pulse criterion and the stability margin determined by the eigenvalues of the system (3.2.13).

For this purpose, in the stability domain of Fig. 3.4, level lines of the maximum eigenvalue modulo z^* were constructed, from which the parameter values λ, ξ_0 were

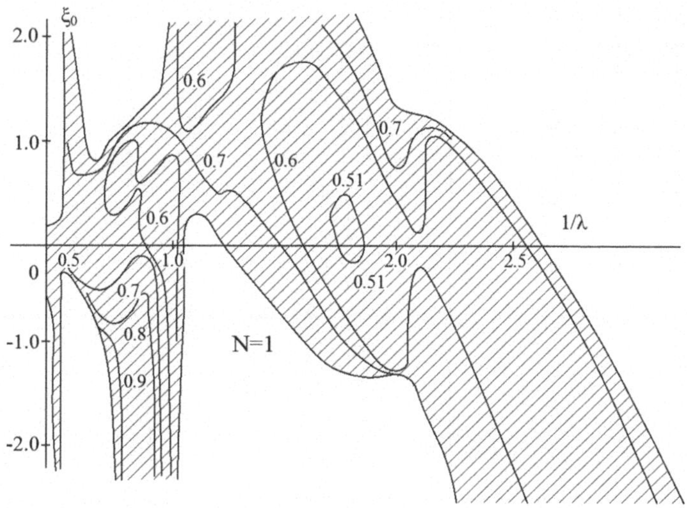

Fig. 3.2 Regions of stability of a single-impact periodic mode of motion of a single-mass mechanism. Lines of the level of the maximum modulo eigenvalue of matrix *A* are presented

Fig. 3.3 Regions of stability of a single-impact periodic mode of motion of a two-mass mechanism. Lines of the level of the maximum modulo eigenvalue of matrix *A* are presented

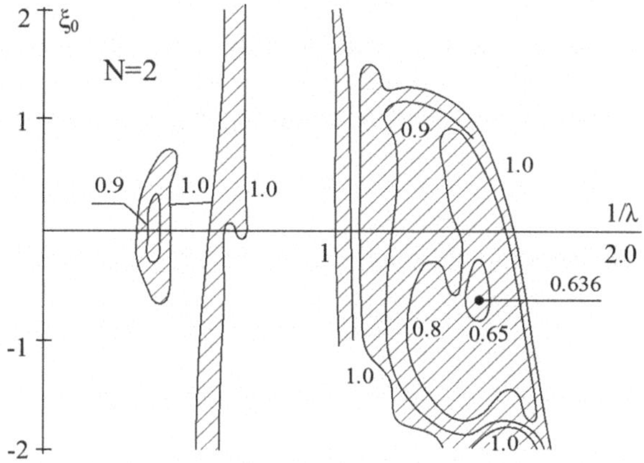

found that satisfy the minimum condition for the maximum eigenvalue of the matrix *A*.

Thus, for $N = 1$, the minimum value z^* is in the vicinity of the point $\lambda^{-1} = 1.77, \xi_0 = 0.08$, for $N = 2$ in $\lambda^{-1} = 1.54, \xi_0 = -0.6$ and for $N = 4$ in $\lambda^{-1} = 1.04, \xi_0 = -1.04$. Hence, with increasing mass number, the optimal point shifts to the region of large λ (smaller ω). Furthermore, if the optimal point for $N = 1$ is far from the resonance, then at $N = 4$, it is in the resonance vicinity. With increasing N, the gap in the negative region is observed to decrease. In order to tune the system to the optimum mode with increasing mass number, it is necessary to increase the preliminary compression of the springs.

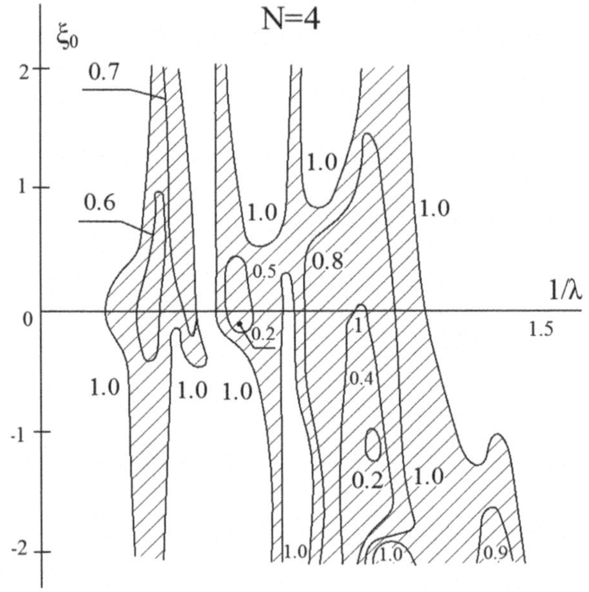

Fig. 3.4 Regions of stability of a single-impact periodic mode of motion of a four-mass mechanism. Lines of the level of the maximum modulo eigenvalue of matrix A are presented

Comment. The study of the stability of a fixed point of a point map can be reduced to the problem whether the roots of a characteristic polynomial belong to a unit circle.

After having varied the point mapping in normal coordinates, we add into (3.2.14)

$$\delta A_n^1 = z\delta A_n^0,\ \delta \dot{A}_n^1 = z\delta \dot{A}_n^0$$

Considering the relation

$$\delta \dot{A}_k^+ = \delta \dot{A}_k^0 - (1+R)a^{-1} \sin \alpha_k N \delta \dot{\xi}_N^0 \tag{3.2.15}$$

we obtain

$$\begin{aligned}
&(z - \cos T\Lambda_n)\delta A_n^0 - \Lambda_n^{-1} \sin T\Lambda_n \\
&\quad = \left\{ (z-1)\frac{1+R}{1-R} g \cos \tau_0 \frac{\sin \alpha_n N}{a} \right. \\
&\quad \left. + \beta_n \cos \tau_0 (z - \cos T\Lambda_n) + \beta_n \frac{\sin T\Lambda_n}{\Lambda_n} \sin \tau_0 \right\} \delta\tau_0 \\
&\quad - (1+R)\Lambda_n^{-1} \sin T\Lambda_n \frac{\sin \alpha_n N}{a} \delta \xi_N^0 \\
&\Lambda_n \sin T\Lambda_n \delta A_n^0 + (z - \cos T\Lambda_n)\delta \dot{A}_n^0 \\
&\quad = \left\{ \frac{1+R}{1-R} g \Lambda_n \cos \tau_0 \times \right.
\end{aligned}$$

$$\times \frac{\sin \alpha_n N}{a}(z-1)ctg\frac{T}{2}\Lambda_n - \beta_n[\sin \tau_0(z - \cos T\Lambda_n)$$
$$- \Lambda_n \sin T\Lambda_n \cos \tau_0\}\delta\tau_0$$
$$- (1+R)\cos T\Lambda_n a^{-1} \sin \alpha_n N \delta\dot{\xi}_N^0 \tag{3.2.16}$$

The relation (3.2.16) is a system of two equations with respect to δA_n^0, $\delta \dot{A}_n^0$, and the solution of which is written in the form

$$\delta A_n^0 = -\frac{(1+R)z}{\Delta n}\frac{\sin T\Lambda_n}{\Lambda_n}\frac{\sin \alpha_n N}{a}\delta\dot{\xi}_N^0$$
$$+ \left\{\beta_n + \frac{(z^2-1)}{\Delta n}\frac{1+R}{1-R}g\frac{\sin \alpha_n N}{a}\right\}\cos \tau_0 \delta\tau_0$$
$$\delta \dot{A}_n^0 = -\frac{(1+R)}{\Delta n}\frac{\sin \alpha_n N}{a}(z\cos T\Lambda_n - 1)\delta\dot{\xi}_N^0 +$$
$$+ \left[-\beta_n \sin \tau_0 + \frac{1+R}{1-R}g\Lambda_n\frac{\cos \tau_0}{\Delta n}\frac{\sin \alpha_n N}{a}(z-1)^2 ctg\frac{T}{2}\Lambda_n\right]\delta\tau_0 \tag{3.2.17}$$

where $\Delta k = z^2 - 2z\cos T\Lambda_n + 1$.

Idealization of the instantaneous impact of lower mass against a fixed limiter using (3.2.17) gives

$$(1+R)z\ell_2(z)\frac{\delta\dot{\xi}_N^0}{\delta\tau_0} = g\left[1 + \frac{1+R}{1-R}(z^2-1)\ell_1(z)\right]\cos \tau_0$$
$$[1 + (1+R)z\ell_3(z) - (1+R)\ell_1(z)]\frac{\delta\dot{\xi}_N^0}{\delta\tau_0}$$
$$= g\left[-\sin \tau_0 + \frac{1+R}{1-R}\cos \tau_0 (z-1)^2 \ell_4(z)\right] \tag{3.2.18}$$

where it is denoted

$$\ell_1(z) = a^{-2}\sum_{k=1}^{N}\frac{\sin^2 \alpha_n N}{\Delta k}$$
$$\ell_2(z) = a^{-2}\sum_{k=1}^{N}\frac{\sin^2 \alpha_n N}{\Delta k}\frac{\sin T\Lambda_k}{\Lambda_k}$$
$$\ell_3(z) = a^{-2}\sum_{k=1}^{N}\frac{\sin^2 \alpha_n N}{\Delta k}\cos T\Lambda_k$$
$$\ell_4(z) = a^{-2}\sum_{k=1}^{N}\frac{\sin^2 \alpha_n N}{\Delta k}\Lambda_k ctg\frac{T}{2}\Lambda_k \tag{3.2.19}$$

From (3.2.18), we obtain the equation with respect to z

$$(1+R)z\ell_2(z)\left[\frac{1+R}{1-R}\cos\tau_0(z-1)^2\ell_4(z)-\sin\tau_0\right]$$
$$-\frac{1-R}{2}\left[1+\frac{1+R}{1-R}(z^2-1)\ell_1(z)\right]^2\cos\tau_0=0$$

from which after having reduced to common denominator a characteristic polynomial of $4N$ degree is obtained. The additional $2N$ of polynomial roots is zeros $\triangle\ n\left(z_n^0=e^{\pm iT\Lambda_n}\right)$. It is easy to show that z_n^0 satisfies (3.2.20). Having found the polynomial roots employing the Mueller method [11], by example, the problem of the stability of a fixed point will be solved when the relations $|z_i|<1$ are held.

Note. We note that formulas (3.2.11), which determine the stability of a fixed point (3.1.38), corresponding to a single-impact n-fold periodic motion, coincide with similar relations in [12, 13] for a single-mass ($N=1$) and two-mass ($N=2$) vibro-impact mechanism. So, passing in the system (3.2.11) from variable k to $N=1$ using the relations

$$\begin{aligned}
&a^2=3/4,\lambda_1=\pi/3,\Lambda_1=\lambda,A_{11}=\cos T\lambda,A_{11}^2=\lambda^{-1}\sin T\lambda\\
&B_1^1=\alpha\cos T\lambda,B_1^2=\lambda^{-1}\alpha\sin T\lambda,B_1^3=\alpha\lambda\sin T\lambda,B_1^4=A_1^6=\alpha\\
&\ell_1^2=\lambda ctg\frac{T}{2}\lambda,\beta_n=\alpha=g=(\lambda^2-1)^{-1}\\
&\Gamma_1^2=\frac{1}{2}\left[(1+R)\lambda ctg\frac{T}{2}\lambda-(1-R)tg\tau_0\right]\\
&\Gamma_1^3=\frac{1}{2}[(1+R)+(1-R)\left(\cos T\lambda-\lambda^{-1}\cos T\lambda tg\tau_0\right]\\
&\Gamma_1^4=\frac{1}{2}[(1+R)\lambda ctg\frac{T}{2}\lambda-(1-R)(\lambda\sin T\lambda+\cos T\lambda tg\tau_0]
\end{aligned}\tag{3.2.20}$$

we obtain

$$\begin{aligned}
\delta\tau_1=&\frac{\alpha}{\dot{\xi}_0}\left\{\cos T\lambda\cos\tau_0-\lambda^{-1}\sin T\lambda\sin\tau_0+\frac{1+R}{1-R}\cos\tau_0\right\}\delta\tau_0\\
&+R\frac{\sin T\lambda}{\lambda}\dot{\xi}_0^{-1}\delta\dot{\xi}_0\\
\delta\dot{\xi}_1=&\left\{\lambda\sin T\lambda\cos\tau_0+\cos T\lambda\sin\tau_0-\frac{1+R}{1-R}\lambda ctg\frac{T}{2}\lambda\cos\tau_0\right.\\
&+\frac{\alpha}{\dot{\xi}_0}\left[\frac{1+R}{1-R}\lambda ctg\frac{T}{2}\lambda\cos\tau_0-\sin\tau_0\right]\left[\cos T\lambda\cos\tau_0-\frac{\sin T\lambda}{\lambda}\sin\tau_0\right.\\
&\left.\left.+\frac{1+R}{1-R}\cos\tau_0\right\}\right]\delta\tau_0-R\alpha\left\{\alpha^{-1}\cos T\lambda-\dot{\xi}_0^{-1}\left[\frac{1+R}{1-R}\lambda ctg\frac{T}{2}\lambda\cos\tau_0\right.\right.\\
&-\sin\tau_0]\lambda^{-1}\sin T\lambda\}\delta\dot{\xi}_0
\end{aligned}\tag{3.2.21}$$

Similar relations can be obtained by varying the point mapping in [12].

References

1. Nagaev, R.F.: Regular impulsive motions in a one-dimensional system. PMM **31**(2) (1967)
2. Gorokhov, V.A.: On periodic motions of a system with impact interaction. Izv. Higher Uch. Head, Radiophys. **9**(2) (1966)
3. Ostapenko, V.A.: Mechanical Vibroimpact Systems. "Scientific thought". Kyiv (1966)
4. Panovko, Ya.G., Gubanova, I.I.: Stability and vibrations of elastic systems. Ed. "Science" (1967)
5. Timoshenko, S.P.: Vibrations in Engineering. Ed. "Science" (1967)
6. Zevin, A.A.: On the stability of periodic motions in the presence of mass collisions. Mach. Sci. **4** (1968)
7. Ryzhik, I.M., Gradshtein I.S.: Tables of integrals, series and products. M-L. 1951
8. Neimark, Yu.I.: The method of point transformations in the theory of nonlinear oscillations. Izv. Higher Uch. Head, Radiophys. **1** (1958)
9. Neimark, Yu.I.: The method of point mappings in the theory of nonlinear oscillations. Izv. Vysh. Uch. Head, Radiophys. **2**(2) (1958)
10. Neimark, Yu.I.: Method of point transformations in the theory of nonlinear oscillations. Izv. Higher Uch. Head, Radiophys. **3**(5–6) (1958)
11. Carloni, R., Sanfelice, R.G., Teel, A.R., Melchiorri, C.: A hybrid control strategy for robust contact detection and force regulation. In: Proceedings of the American Control Conference, pp. 1461–1466, New York City, USA (2007)
12. Bespalova, L.V.: To the theory of vibro-impact mechanism. Izv. Acad. Sci. USSR, OTN, No. 5 (1957)
13. Bespalova, L.V., Metrikin, V.S.: On the theory of a two-mass model of a vibrohammer. Izv. Higher Uch. Head, Radiophys. **9**(2) (1966)

Chapter 4
Investigating the Effect of Viscous Friction Forces on Stability of Vibro-impact Systems

Abstract This chapter is devoted to the study of the influence of viscous friction forces on the stable operation of vibration-impact systems. In § 4.1, the dynamics of a single-mass vibrating impactor are considered taking into account viscous friction forces. The coordinates of fixed points corresponding to single-impact n-fold periodic motions are obtained, and their stability is investigated. The bifurcation parameters of the N- and C-surfaces are found and constructed in space. It is shown that taking into account, the forces of viscous friction lead to smoothing out the infinite discontinuities of individual parts of the stability regions. The magnitude of the impact velocity decreases. Some new properties of bifurcation C-surfaces are indicated, consisting in the fact that the usually existing superposition of single-impact periodic movements with different multiplicities ($n = 1, 2, 3, \ldots$) is eliminated by these bifurcation surfaces. Recommendations are given for choosing optimal parameter values. In the case of resonance, a complete partition of the three-dimensional parameter space into stability regions of single-impact n-multiple periodic motions has been carried out. It is shown that in this case, multiple ($n \geq 2$) stable single-impact periodic movements do not exist. The second paragraph examines the stability of the operation of a two-mass vibration-impact mechanism taking into account viscous friction forces. Finding the coordinates of fixed points, drawing up a characteristic polynomial and studying stability is carried out using a developed special software product. The main conclusions are presented at the end of the chapter.

Keywords Mathematical model · Vibration-impact mechanism · Stability · Fixed point · Point mapping method · Friction forces

The motion of real mechanical systems is always accompanied by friction forces. Therefore, in order to bring the analytical solutions of vibro-impact systems into better compliance with the real conditions, it is necessary to account for the inelastic resistance forces (damping) arising due to friction between dry friction surfaces, air or fluid resistance, internal friction as a result of imperfect elasticity, etc.

The friction force between dry surfaces is calculated using Coulomb's law [1]. According to it, the friction force F is proportional to the normal pressure N acting

V. Metrikin and I. Leonid, *Forced Oscillations of Multidimensional Highly Nonlinear Systems*, Advanced Structured Materials 222,
https://doi.org/10.1007/978-981-97-6779-3_4

between the surfaces,

$$F = \mu N$$

where μ is friction coefficient, the value of which depends on the material of the rubbing bodies and roughness degree of their surfaces.

During friction of lubricated surfaces, the friction force does not depend on the materials that are in contact, but is a function of the motion velocity and lubricant viscosity. In the case of perfectly lubricated surfaces when there is a continuous lubricant film between them, it can be assumed that the friction forces are proportional to the lubricant viscosity and motion velocity. In the case when the body moves in a viscous fluid at a low velocity or when a moving body pushes the fluid through narrow channels, we can assume that the resistance force is proportional to the motion velocity. Later on in our investigation on resistance forces of a complex nature, for simplicity of analysis, we replace them with equivalent viscous resistance proportional to velocity as in [2]. This equivalent resistance should cause the same energy dissipation in one cycle as the actual resistance force.

4.1 Dynamics of a Single-Mass Striker Subject to Viscous Friction

Figure 4.1 presents the vibro-impact model consisting of an elastic (with a stiffness coefficient k) mass M subject to external sinusoidal force $F \sin \omega t$.

Moreover, we believe that the system has viscous friction proportional to the velocity. The dynamics of a single-mass vibro-striker without viscous friction was investigated by L. V. Bespalova in [3].

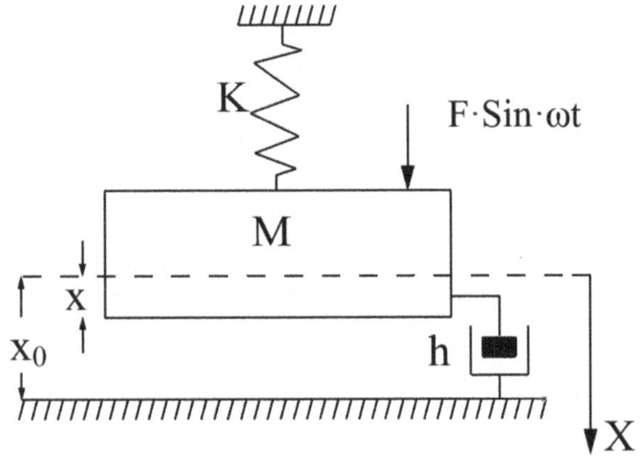

Fig. 4.1 Physical diagram of a vibration-impact mechanism taking into account friction

Having selected the position of the system in which the springs are undeformed, the equations of motion in the intervals between impacts $x < x_0$ will be written in the form

$$M\ddot{x} + h\dot{x} + kx = P + F\sin\omega t \tag{4.1.1}$$

where P is constant force, acting upon the impact mass and $P \neq mg$. F, ω are amplitude and frequency of the driving force, h is viscous friction coefficient. At $x = x_0$, there is an instant impact of mass M against a stationary limiter, characterized by a coefficient of restitution of the impact velocity R. If we denote by $\dot{x}_1$ and $\dot{x}_0$ the velocities immediately after and before the impact, then the relation $x = x_0$ will be held

$$\dot{x}_1 = -R\dot{x}_0 \tag{4.1.2}$$

In dimensionless variables

$$\xi = \frac{M\omega^2}{F}\left(x - k^{-1}P\right),\ \omega t = \tau \tag{4.1.3}$$

Equations (4.1.1) and (4.1.2) will be rewritten in the form

$$\ddot{\xi} + 2\mu\dot{\xi} + \lambda^2\xi = \sin\tau \text{ at } \xi < \xi_0 \tag{4.1.4}$$

$$\dot{\xi}_1 = -R\dot{\xi}_0 \text{ at } \xi = \xi_0 \tag{4.1.5}$$

where dimensionless parameters denote

$$2\mu = \frac{h}{M\omega},\ \lambda^2 = \frac{k}{M\omega^2} \tag{4.1.6}$$

The system is characterized by four dimensionless parameters λ, μ, R, ξ_0.

Since the phase space of a system with one degree of freedom is non-autonomous, then it is three-dimensional in coordinates $\xi, \dot{\xi}, \tau$. When studying the system, it is convenient to represent the phase space in a cylindrical coordinate system θ, r, z, as in [3], where $r = r_0 + \xi_0 - \xi, \theta = \tau, z = \dot{\xi}$ (on further statement, we retain the previous notations $\tau, \dot{\xi}$ for coordinates θ, z). Thereafter, we obtain that the limiter plane $\xi = \xi_0$ in the phase space r, θ, z corresponds the surface Π of radius cylinder $r = r_0$. And the space G outside the cylinder corresponds to domain above the surface of the oscillation limiter $\xi < \xi_0$.The space inside the cylinder corresponds to the domain below the limiter plane $\xi > \xi_0$.

In the case of impact motions of the system at $R \neq 0$, the mapping point can move as follows: while moving in G domain, it hits Π surface at some point $A_0(\tau_0, \dot{\xi}_0)$, then it instantly moves along the generatrix of the cylinder to the point $A_0'(\tau_0, -R\dot{\xi}_0)$, whereafter it moves in the subspace G until it hits again the surface Π at some point

$A_1(\tau_1, \dot{\xi}_1)$. Consequently, in this case, the problem is reduced to a point mapping of the cylinder surface into itself (we denote it by the S-mapping).

On the surface Π, in addition to the trajectories parallel to the generatrix of the cylinder, the part of the circle $\dot{\xi} = 0$ can also be a trajectory. If the striker hits the limit surface ($\dot{\xi}_1 = 0$) and its acceleration is either greater than or equal to zero ($\ddot{\xi} \geq 0$), then the image point will circle until acceleration is negative ($\ddot{\xi} < 0$). This part of the trajectory corresponds to a long stop of the striker.

In the case of an absolutely inelastic impact during the motion of the system, the image point, after hitting any point on the cylindrical surface instantly moves along the generatrix of the cylinder to the circle $r = r_0, \dot{\xi} = 0$. Then if acceleration has negative sign, it again moves to region G, whereas at positive acceleration, the image point moves along the indicated circle until the acceleration changes the sign. Thus, in the case of an absolutely inelastic impact, the image point will be hitting each time the circle $r = r_0$ in the plane $\dot{\xi} = 0$, and therefore, investigation of its motion can be reduced to studying the point mapping of the circle into itself.

Let us consider an elastic continuous impact ($R \neq 0$).

The motion of the image point in case of oscillatory motions of the system ($\lambda > \mu$) in subspace G, passing through the point A_0', is described by the equations

$$\begin{aligned}\xi(\tau) &= \rho \sin\tau - \zeta \cos\tau + e^{-\mu(\tau-\tau_0)}\left[A\alpha^{-1}\sin\alpha(\tau-\tau_0)\right.\\&\quad \left.+B\cos\alpha(\tau-\tau_0)\right]\\ \dot{\xi}(\tau) &= \rho\cos\tau + \zeta\sin\tau + e^{-\mu(\tau-\tau_0)}\left[(A-\mu B)\cos\alpha(\tau-\tau_0)\right.\\&\quad \left.-\left(\mu\alpha^{-1}A + \alpha B\right)\sin\alpha(\tau-\tau_0)\right]\end{aligned} \tag{4.1.7}$$

where the following notations are introduced

$$\begin{aligned}A &= \left[\xi_0\mu - R\dot{\xi}_0 + (\mu\zeta - \rho)\cos\tau_0 - (\zeta + \mu\rho)\sin\tau_0\right]\\ B &= [\xi_0 - \rho\sin\tau_0 + \zeta\cos\tau_0],\ \ \Delta = (\lambda^2-1)^2 + 4\mu^2\\ \alpha &= \sqrt{\lambda^2 - \mu^2},\ \ \rho = \left(\lambda^2 - 1\right)\Delta^{-1},\ \ \zeta = 2\mu\Delta^{-1}\end{aligned} \tag{4.1.8}$$

With increasing time τ, starting from τ_0, the image point having emerged from the surface Π, after a certain period of time hits it against time τ_1. Time τ_1 is derived from the first Eq. (4.1.7), assuming that $\xi = \xi_0$, striker velocity $\dot{\xi}_1$ at this time is determined by the second Eq. (4.1.7) by substituting $\tau = \tau_1$ into it, obtained from the first equation. The value τ_1 must be greater than τ_0 and the nearest to it.

Thus, a point mapping of a cylindrical surface Π into itself can be written as follows

$$\begin{aligned}\xi_0 &= \rho\sin\tau_1 - \zeta\cos\tau_1 + e^{-\mu(\tau_1-\tau_0)}\left[A\alpha^{-1}\sin\alpha(\tau_1-\tau_0)\right.\\&\quad\left.+B\cos\alpha(\tau_1-\tau_0)\right]\\ \dot{\xi}_1 &= \rho\cos\tau_1 + \zeta\sin\tau_1 + e^{-\mu(\tau_1-\tau_0)}\left[(A-\mu B)\cos\alpha(\tau_1-\tau_0)\right.\\&\quad\left.-\left(\mu\alpha^{-1} + \alpha B\right)\sin\alpha(\tau_1-\tau_0)\right]\end{aligned} \tag{4.1.9}$$

A closed path in the phase space corresponds to periodic motion of the striker, since the image point returns to its initial position in a motion period. It is clear from the phase space structure that the trajectory can be closed only when it makes one or several circles around the cylinder. Hence, the periodic motion of the system can only be with a period that is a multiple of the driving force period $T = 2\pi n$, where $n = 1, 2, \ldots$. The search form–impact n-multiple periodic motion is reduced to finding a fixed point S^m-mapping (mis the number of repeated S-mapping) which contains $2(m+1)$ equations of the form

$$\begin{aligned}
\xi_0 &= \rho \sin \tau_i - \zeta \cos \tau_i + \mathrm{e}^{-\mu(\tau_i - \tau_{i-1})}\big[A\alpha^{-1} \sin \alpha(\tau_i - \tau_{i-1}) \\
&\quad + B \cos \alpha(\tau_i - \tau_{i-1})\big] \\
\dot{\xi}_i &= \rho \cos \tau_i + \zeta \sin \tau_i + \mathrm{e}^{-\mu(\tau_i - \tau_{i-1})}\big[(A - \mu B) \cos \alpha(\tau_i - \tau_{i-1}) \\
&\quad - \left(\mu\alpha^{-1} + \alpha B\right) \sin \alpha(\tau_i - \tau_{i-1})\big] \\
\dot{\xi}_m &= \dot{\xi}_0, \ \ \tau_m - \tau_0 = 2\pi n, \ \ (i = 1, 2, \ldots, m)
\end{aligned} \tag{4.1.10}$$

To find the coordinates of a fixed point of a single-impact n-fold periodic mode, it is necessary to add $m = 1, \ \tau_1 - \tau_0 = 2\pi n, \ \dot{\xi}_1 = \dot{\xi}_0$ into (4.1.10), then we obtain

$$\begin{aligned}
\xi_0 &= \rho \sin \tau_0 - \zeta \cos \tau_0 + \mathrm{e}^{-\mu T}\left(A\alpha^{-1} \sin \alpha T + B \cos \alpha T\right) \\
\dot{\xi}_0 &= \rho \cos \tau_0 + \zeta \sin \tau_0 + \mathrm{e}^{-\mu T}[(A - \mu B) \cos \alpha T \\
&\quad - \left(\mu\alpha^{-1} + \alpha B\right) \sin \alpha T\big]
\end{aligned} \tag{4.1.11}$$

Solving this system with respect to $\sin \tau_0$ and $\cos \tau_0$, we will have

$$\begin{aligned}
\sin \tau_0 &= p^{-1} q^{-2}\left\{p\rho\xi_0 + \left[\zeta g - \mathrm{e}^{-d}(f(\mu\zeta - \rho) + \zeta\varsigma)\right]\dot{\xi}_0\right\} \\
\cos \tau_0 &= p^{-1} q^{-2}\left\{-p\zeta\xi_0 + \left[\rho g - \mathrm{e}^{-d}(f(\mu\rho + \zeta) + \rho\varsigma)\right]\dot{\xi}_0\right\}
\end{aligned} \tag{4.1.12}$$

From Eq. (4.1.12) for impact velocity, we derive the following square equation

$$\left[(g - \mathrm{e}^{-d} F)^2 + \mathrm{e}^{-2d} f\right]\dot{\xi}_0^2 + 2fp\mathrm{e}^{-d}\xi_0\dot{\xi}_0 + \left(\xi_0^2 - q^2\right)p^2 = 0 \tag{4.1.13}$$

where it is denoted

$$\begin{aligned}
F &= f\mu + \varsigma, f = \alpha^{-1}(1 + R) \sin \alpha T, \ \ \varsigma = (1 - R) \cos \alpha T \\
P &= \left(\mathrm{e}^{-2d} - 2\mathrm{e}^{-d} \cos \alpha T + 1\right), \ \ g = \left(1 - \mathrm{Re}^{-2d}\right) \\
q^2 &= \zeta^2 + p^2, \ \ d = \mu T
\end{aligned} \tag{4.1.14}$$

Note Eqs. (4.1.12) and (4.1.13) at $\mu = 0$ coincide with the relative equations in [3].

Due to adopted coordinate system, the impact occurs at positive values of impact mass velocity ($\dot{\xi}_0 \geq 0$). Therefore, the boundary of positivity of impact velocity at

$f\xi_0 > 0$ is

$$|\xi_0| = q \tag{4.1.15}$$

At $f\xi_0 < 0$, the boundary has the form

$$|\xi_0| = q\sqrt{1 + \mathrm{e}^{-2d}f^2(g + F\mathrm{e}^{-d})^{-2}} \tag{4.1.16}$$

A periodic single-impact n-fold motion will exist with continuous parameter change until this motion is stable or until an additional impact occurs for the period of motion. In the first case, the stability boundary N is determined by the stability of the fixed point of the point mappings and can be found [4–6] as boundaries N_+, N_-, N_φ. In the second case, the disappearance of a single-impact periodic motion is caused [4–6] by the discontinuity of the point mapping S. These bifurcation boundaries [7] are denoted by C_τ.

For stability, as in the preceding sections, a characteristic equation $\sum_{i=0}^{2} A_i z^{2-i} = 0$ is compiled as a result of linearization of S-mapping equations in a neighborhood of a fixed point (4.1.12) and (4.1.13), assuming then $\delta\tau_1 = z\delta\tau_0$ and $\delta\dot{\xi}_1 = z\delta\dot{\xi}_0$. After transformation series, the characteristic equation is written as

$$\begin{aligned} &\dot{\xi}_0 z^2 + \left\{[N\cos\tau_0 - M\sin\tau_0 + L\dot{\xi}_0]\sin\alpha T + 2R\dot{\xi}_0\cos\alpha T\right\} \\ &\quad\times \mathrm{e}^{-d}z + R^2\mathrm{e}^{-2d}\dot{\xi}_0 = 0 \end{aligned} \tag{4.1.17}$$

where equations N, M, L have the form

$$\begin{aligned} N &= \alpha^{-1}\left[\varsigma\left(\lambda^2 - 1\right) - 2\mu q\right] \\ N &= \alpha^{-1}\left[\varsigma\left(\lambda^2 - 1\right) - 2\mu q\right] \\ L &= (1 + R)\alpha^{-1}\lambda^2 \end{aligned} \tag{4.1.18}$$

For stability of a fixed point, it is necessary and sufficient that the following three inequalities hold simultaneously [8]:

$$\sum_{i=0}^{2} A_i > 0,\ \sum_{i=0}^{2}(-1)^i A_i > 0,\ A_0 - A_2 > 0 \tag{4.1.19}$$

The last inequality (the region limited by the surface N_φ) is reduced to the condition $\left(1 - R^2\mathrm{e}^{-2d}\right)\dot{\xi}_0 > 0$. Since the impact velocity is always positive and $0 < R < 1, 0 < \mathrm{e}^{-2d} < 1$ and due to the fact that $d = T\mu > 0$, this inequality always holds. Therefore, N_φ boundary doesn't exist in the stability region. It is directly seen from (4.1.12) and (4.1.13) that for the same values of the system parameters, there are two fixed points of S-mapping $A_1\left(\tau_0, \dot{\xi}_{01}\right), A_2\left(\tau_0, \dot{\xi}_{02}\right)$, where values $\dot{\xi}_{01}, \dot{\xi}_{02}$ are determined from the equality

$$\dot{\xi}_{0i} = p\frac{-f e^{-d}\xi_0 \pm \sqrt{q^2 e^{-2d} f^2 + (q^2 - \xi_0^2)(g - e^{-d}F)^2}}{(g - e^{-d}F)^2 + e^{-2d} f^2}, \quad i = 1, 2 \qquad (4.1.20)$$

And the value $\dot{\xi}_{01}$ corresponds to the value $\dot{\xi}_0$ with a plus sign in front of the root, whereas the value $\dot{\xi}_{02}$ to that with a minus sign.

It is shown that periodic motion corresponding to a fixed point A_2 for any set of parameters is unstable. For this, we substitute the first inequality (4.1.19) with the coordinates of the fixed point, where the value $\dot{\xi}_{02}$ is impact velocity. After some rather cumbersome transformations, we can obtain the inequality

$$-\sqrt{q^2 e^{-2d} f^2 + (q^2 - \xi_0^2)(g - e^{-d}F)^2} > 0$$

which is not held at any values of system parameters R, μ, λ, ξ_0. Therefore, the periodic motion corresponding to a fixed point $A_2(\tau_0, \dot{\xi}_{20})$ is always unstable.

Thus, the stability region of a single-impact n-fold periodic motion is limited by two surfaces N_+, N_-. The boundaries of stability N_+, which coincide with the boundaries of existence, have the form (4.1.15), (4.1.16). The equation defining the boundary N_- can be written as

$$\dot{\xi}_0\left(1 + R^2 e^{-2d}\right) = \left[(N \cos\tau_0 - M \sin\tau_0 + L\xi_0)\sin\alpha T + 2R\dot{\xi}_0 \cos\alpha T\right]e^{-d} \qquad (4.1.21)$$

In addition to the loss of stability during a continuous change in the system parameters, a single-impact n-fold periodic motion can turn into a more complex two, three, etc., impact motion. The bifurcation boundary of the transformation of a single-impact into a two-impact is determined by the conditions

$$\begin{aligned} &\xi(\tau^*) = \xi_0, \dot{\xi}(\tau^*) = 0, \ \tau_0 < \tau^* < \tau_1 \\ &\tau_1 - \tau_0 = T \end{aligned} \qquad (4.1.22)$$

To construct the bifurcation curve (4.1.22), by excluding τ^* from two equations $\xi(\tau^*) = \xi_0, \dot{\xi}(\tau^*) = 0$, we obtain a system of two transcendental equations with respect to the system parameters the solutions of which give the required boundary surface in the parameter space. However, due to the specification of system (4.1.22), the problem of constructing a bifurcation surface can be simplified by substituting the coordinates of the fixed point (4.1.12), (4.1.13) into Eqs. (4.1.22). As a result, a transcendental equation independent of ξ_0 form is obtained

$$E_1E_4 - E_2E_3 = 0 \qquad (4.1.23)$$

from which, at fixed parameter values, the moment τ^* is derived and the reduced gap is determined from the relation

$$\xi_0^2\left[(E_1E_2^{-1}M^{-1} - \varsigma')^2 + M'\right] = N' \tag{4.1.24}$$

where

$$\begin{aligned}
&E_1 = Q_1 + Q_2T_1 + Q_3T_2, E_2 = Q_2T_3 + Q_3T_4 - Q_4\\
&E_3 = P_1 + P_2T_1 + P_3T_2, E_4 = P_2T_3 + P_3T_4 - P_4\\
&Q_1 = \Pi_0 - \mathrm{e}^{\mu\beta}, \Pi_0\mu\alpha^{-1}\sin\alpha\beta + \cos\alpha\beta\\
&\Pi_1 = \mu\alpha^{-1}\sin\alpha\beta - \cos\alpha\beta,\\
&Q_2 = [\Pi_4 - \rho\cos\alpha\beta - \alpha^{-1}(\mu\rho + \zeta)\sin\alpha\beta],\\
&\Pi_2 = \alpha\sin\alpha\beta + \mu\cos\alpha\beta,\\
&Q_3 = [\Pi_3 + \zeta\cos\alpha\beta + \alpha^{-1}(\mu\zeta - \rho)\sin\alpha\beta],\\
&\Pi_3 = \mathrm{e}^{\mu\beta}[\rho\sin\beta - \zeta\cos\beta),\\
&Q_4 = R\alpha^{-1}\sin\alpha\beta, \Pi_4 = \mathrm{e}^{\mu\beta}[\rho\cos\beta + \zeta\sin\beta),\\
&P_1 = \mu\Pi_1 - \Pi_2, P_2 = [\Pi_1(\eta\rho + \zeta) + \Pi_2\rho - \Pi_3]\\
&P_3 = [\Pi_4 - \Pi_1(\mu\zeta - \rho) - \Pi_2\zeta], P_4 = -R\Pi_1, M' = \left(g - \mathrm{e}^{-d}F\right),\\
&\varsigma' = f\mathrm{e}^{-d}, N' = q^2\left(\varsigma'^2 + M'\right), M = Pq^2/N', \tau^* - \tau_0 = \beta
\end{aligned} \tag{4.1.25}$$

From Eq. (4.1.23), for given λ at fixed R and μ we find β, while ξ_0 is obtained from Eq. (4.1.24) for λ and found β from (4.1.23). Equation (4.1.23) in the interval from 0 to 2π can have N roots $\beta_1, \beta_2, \ldots, \beta_N = \{\beta_k\}$. Of all the roots β_k one should choose the root $\beta^* = \beta_{ki}$, where $1 \le i \le N$, to which the largest value of the reduced gap ξ_0 corresponds, i.e.,

$$\max_k|\xi_0(\beta_k, \lambda, \mu, R)| = \left|\xi_0\left(\beta^*, \lambda, \mu, R\right)\right| \tag{4.1.26}$$

However, from Eq. (4.1.24), we can find only the absolute value of the reduced gap. To find the true sign ξ_0, an oscillogram of the impact mass motion is constructed based on differential equations (4.1.4) and (4.1.5), corresponding to at least one set of system parameters, in which the gap is smaller than the absolute value $|\xi_0(\beta^*, \lambda, \mu, R)|$ derived from (4.1.24). If at the same time it turns out that for this set of parameters, there is a periodic single-impact motion, then it is necessary to choose a minus sign at the gap. Then a single-impact motion a turn into multi-impact only with gap reducing. If a single-impact motion does not exist, then for this set of parameters, there exist m-impact ($m > 1$) n-fold periodic modes and therefore, the positive gap value is chosen, equal to $|\xi_0(\beta^*, \lambda, \mu, R)|$. With increasing gap, multi-impact motions can turn into single ones.

Having constructed the bifurcation curve in such a way, we can conclude that a single-impact periodic motion will exist and be stable for the gap value larger than the found ξ_0 value.

Using the formulas obtained above, the space of the system parameters was partitioned into stability regions.

Figures 4.2, 4.3 and 4.4 present in the plane $1/\lambda$, ξ_0 the regions of N-stability (hatching from left to right) of single-impact periodic modes of different multiplicities $n = 1, 2, 3$, for the fixed values of two other parameters of the restitution coefficient equal to 0.6 and reduced damping coefficient $\mu = 0.05$. Comparison of these graphs with similar ones without damping [3] shows that the presence of viscous friction in the system leads to joining of separate parts of stability regions due to smoothing of infinite boundary discontinuities.

The dashed line in these figures corresponds to the bifurcation boundary C_τ. The dash-dotted line shows the boundary of impact less motions. A single-impact periodic motion occurs only in regions with double hatching. The most significant parts of these regions are located for $n = 1, 2, 3$ at values λ^{-1} slightly larger than 1, 3, 5, respectively. The calculation results showed that for any n, the widest parts of the stability regions lie for the values λ, determined from the formula

$$\lambda_n^{-1} > (2n - 1), \quad (n = 1, 2, 3...) \tag{4.1.27}$$

To study in detail the effect of the damping coefficient of stability region, Figs. 4.5, 4.6, 4.7, 4.8 and 4.9 show the partition of the parameter plane $1/\lambda$, ξ_0 into stability regions of a single-impact periodic motion for values μ, equal to 0.001, 0.01, 0.1, 0.25, and 0.5, the coefficient of restitution remaining constant, equal to 0.6.

It can be seen from these figures that with μ growth, the boundary of stability regions become smoother and the regions narrow significantly with respect to ξ_0.

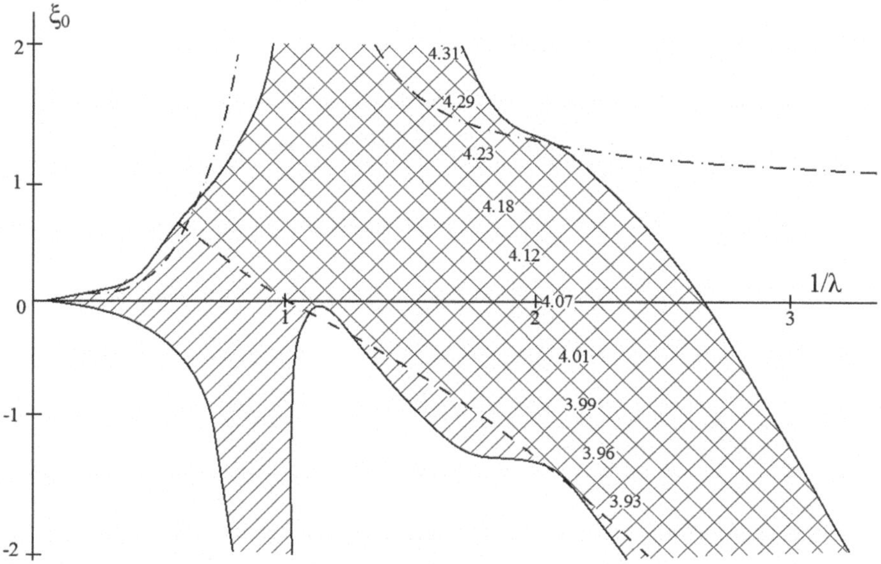

Fig. 4.2 Regions N-stability (shading from left to right) of single-impact periodic motion modes for $n = 1$, $R = 0.6$, $\mu = 0.05$

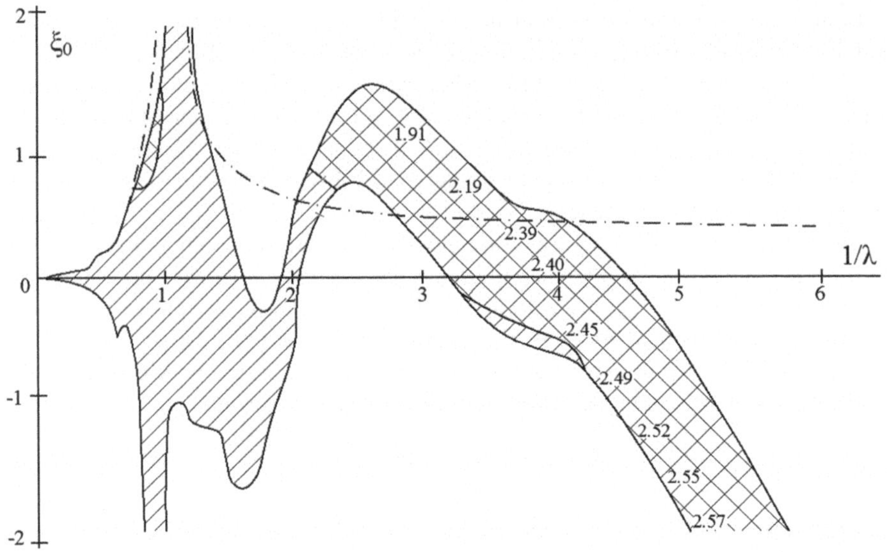

Fig. 4.3 Regions N-stability (shading from left to right) of single-impact periodic motion modes for $n = 2$, $R = 0.6$, $\mu = 0.05$

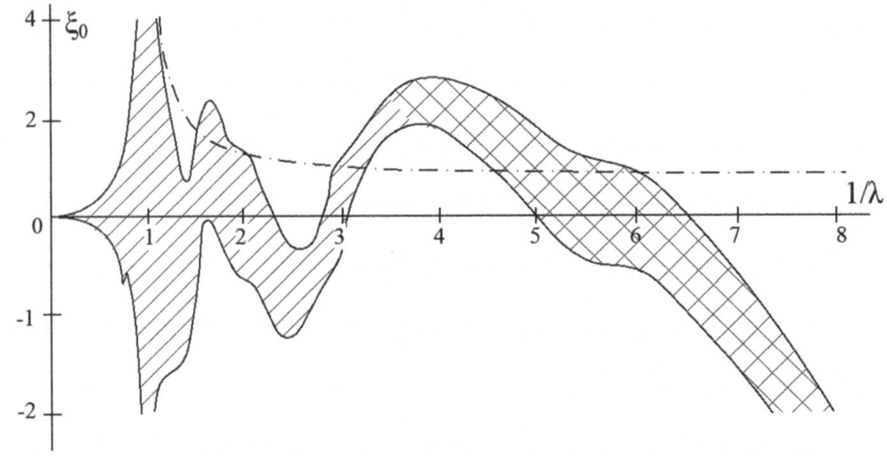

Fig. 4.4 Regions N-stability (shading from left to right) of single-impact periodic motion modes for $n = 3$, $R = 0.6$, $\mu = 0.05$

Consequently, at a large energy dispersion in the system between impacts, the stable motion occurs in a rather narrow part of the plane $1/\lambda$, ξ_0. With increasing μ, the stability regions shift to the regions of smaller λ^{-1}, where oscillatory motion of the system is possible. We note some characteristic properties of the bifurcation surface C_τ.

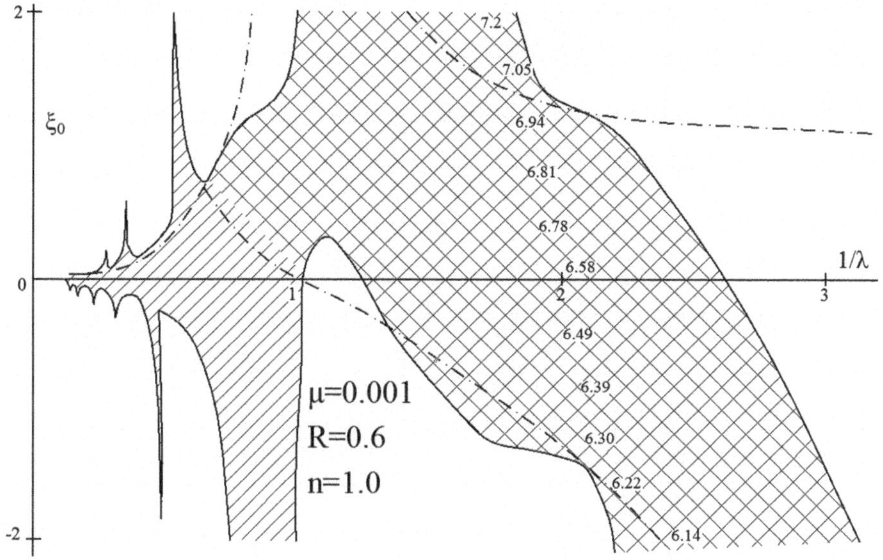

Fig. 4.5 Stability regions (double hatching) of single-impact single periodic motion for $n = 1$, $R = 0.6$, $\mu = 0.001$. The numbers indicate the values of the maximum impact speed at a fixed gap value

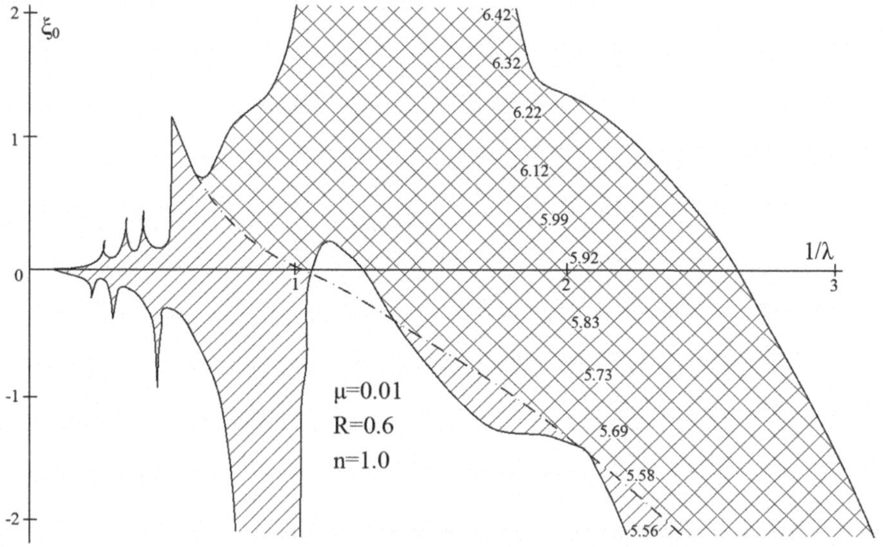

Fig. 4.6 Stability regions (double hatching) of single-impact single periodic motion for $n = 1$, $R = 0.6$, $\mu = 0.01$. The numbers indicate the values of the maximum impact speed at a fixed gap value

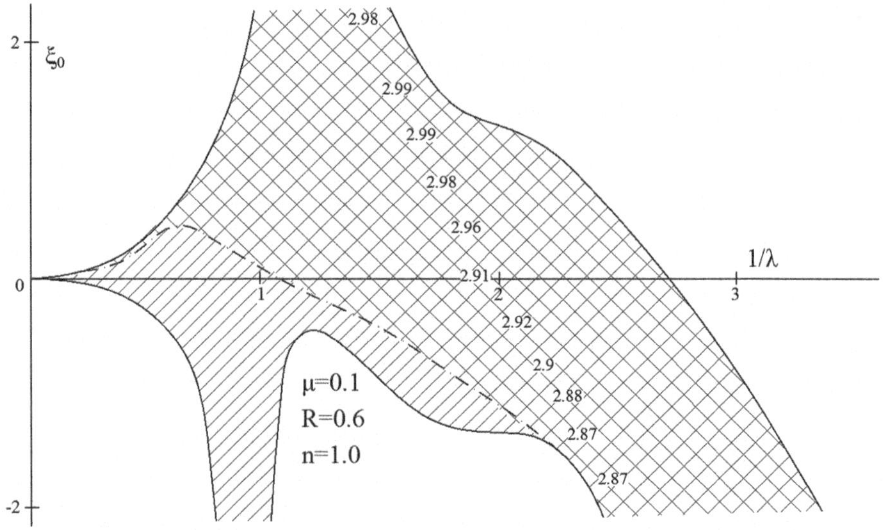

Fig. 4.7 Stability regions (double hatching) of single-impact single periodic motion for $n = 1$, $R = 0.6$, $\mu = 0.1$. The numbers indicate the values of the maximum impact speed at a fixed gap value

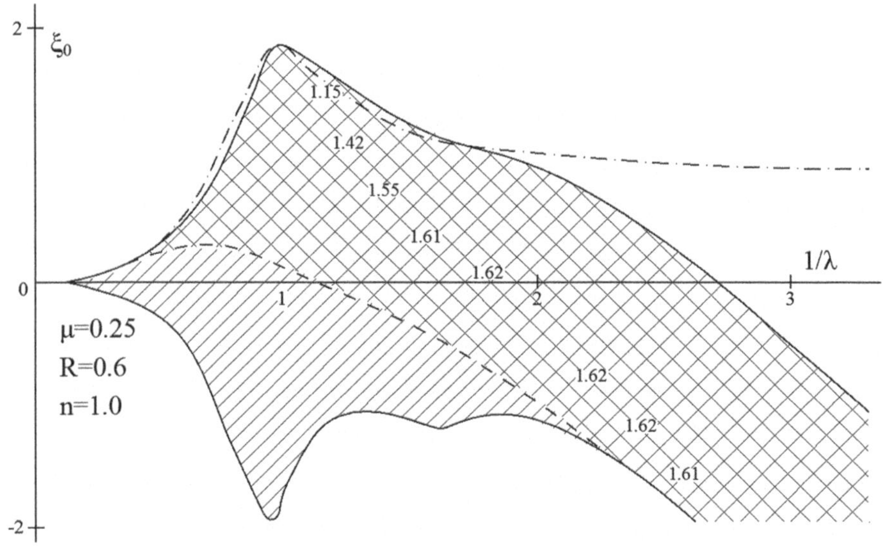

Fig. 4.8 Stability regions (double hatching) of single-impact single periodic motion for $n = 1$, $R = 0.6$, $\mu = 0.25$. The numbers indicate the values of the maximum impact speed at a fixed gap value

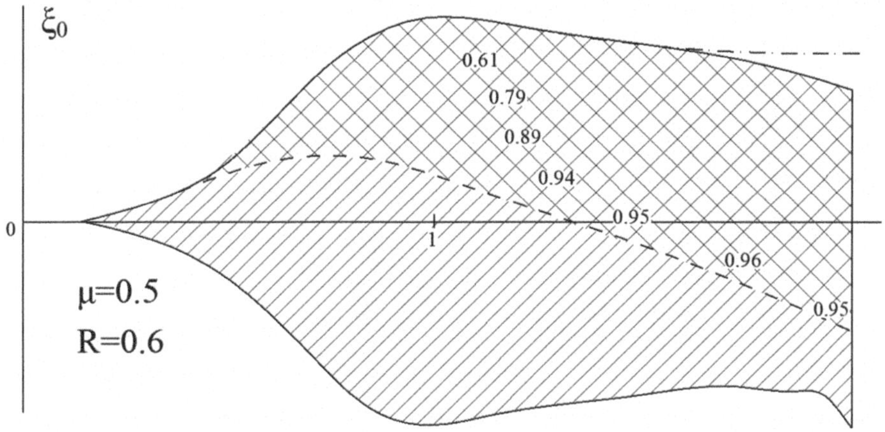

Fig. 4.9 Stability regions (double hatching) of single-impact single periodic motion for $n = 1$, $R = 0.6$, $\mu = 0.5$. The numbers indicate the values of the maximum impact speed at a fixed gap value

First, the boundary C_τ cuts off a significant part of the stability region N (see Figs. 4.2, 4.3, 4.4, 4.5, 4.6, 4.7, 4.8 and 4.9). It follows that to construct this bifurcation surface is as important in investigation into dynamic behavior of vibro-impact systems as N-stability surfaces. The knowledge of location of N-stability surfaces alone is not enough to determine the parameter values for which a stable periodic mode do exists.

Second, when studying the dynamics of vibro-impact systems, the superposition of periodic modes of different multiplicity factor $n = 1, 2, 3$ is observed (see, e.g., [9–14] and other works), i.e., for the same parameter values, there can exist periodic modes both with 2π period and 4π, 6π periods, etc.

However, of single-impact ($m = 1$) n-fold periodic modes showed that in the parameter plane $1/\lambda$, ξ_0 two other parameters when constructing a bifurcation surface, the superposition of regimes with different multiplicity factor disappears (Fig. 4.6). This means that for the same parameter, there can exist periodic motions either with $n = 1$ or with $n = 2$, etc.

However, the absence of superposition of absence of superposition of modes with different multiplicity factor, observed in the plane $1/\lambda$, ξ_0 for different set of parameters R and μ, was not proved in the entire 4D space. Therefore, it is possible that in some other plane section of the parameter space, the superposition of modes with different multiplicity factor will exist.

Thus, Fig. 4.10 presents a location graph of the existence and stability region of single-impact periodic modes, which is a superposition of Figs. 4.2, 4.3 and 4.4. The figure also shows the corresponding graphs for $\mu = 0$. The described above marked properties of C_τ boundaries are clearly emerge from this figure.

It is seen from Fig. 4.10 that for positive gaps ξ_0, the stability regions decrease with increasing damping coefficient μ, whereas for negative gaps, they expand.

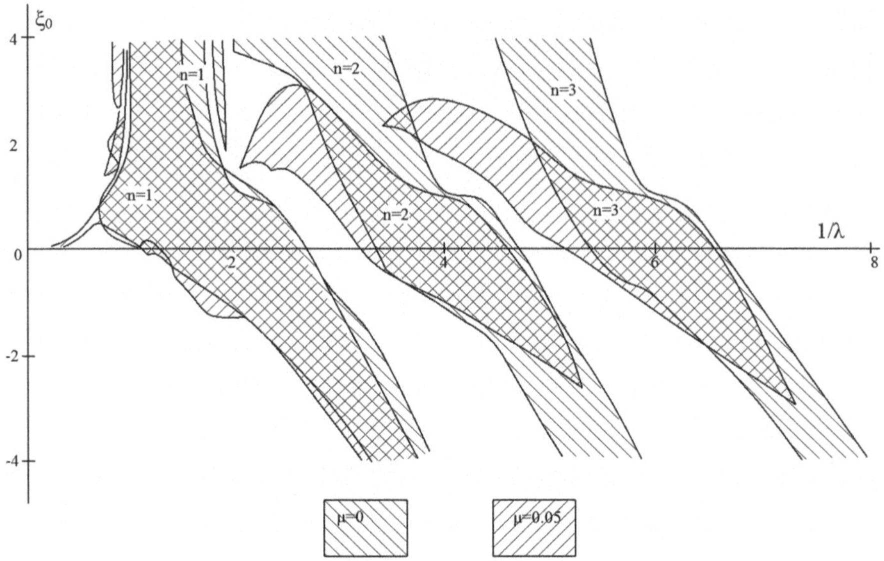

Fig. 4.10 Regions of existence and stability of single-shock periodic regimes

This can also be explained by the following: the friction force reduces the oscillations amplitude of the system, therefore, for positive gaps, when there is a certain gap (clearance) between the striker and the limiter plane, the stability regions are located in regions with smaller gaps as compared with the regions without damping.

The value of the impact velocity as a function of system parameters was studied in the obtained domains of stable single-impact motion. In Figs. 4.2, 4.3, 4.4, 4.5, 4.6, 4.7, 4.8 and 4.9, the number syndicates the maximum impact velocity. The line on which these maximum impact velocity values are located for small values of $\mu(\mu < 0.05)$, extend in regions along λ^{-1}, greater than 1, 3, 5, and intersect the axis $\xi_0 = 0$ for values λ^{-1}, equal to or slightly greater than 2, 4, 6 for $n = 1, 2, 3$, respectively. At high viscous friction forces ($\mu > 0.05$), the maximum line shifts to the region of larger λ. Therefore, for small $\mu(\mu \leq 0.05)$, it is possible to indicate the points on the axis $\xi_0 = 0$, in the neighborhood of which the maximum value $\dot{\xi}_0$ is located. These points are derived from the condition

$$\lambda_n^{-1} = 2n(n = 1, 2, 3, \ldots) \tag{4.1.28}$$

where n is multiplicity factor of oscillation period to the period of external force.

Studies have shown that impact velocity in the stability regions reaches its maximum value for periodic motion with 2π period. Similar results were obtained for other idealizations and by other researchers [13, 15–17].

When tuning the system to the optimal mode, the maximum value of the impact velocity is desirable to be "far" from the boundary of the stability region. The studies have shown that for small gap $\xi_0(|\xi_0| < 1)$, the maximum impact velocity is located in the middle of the stability region and for gaps when $|\xi_0| > 1$, it approaches the region boundaries and for positive values to the right boundary while for negative ones to the left. Therefore, the following recommendations for tuning the system to the optimal mode may be given: when tuning the gap to the optimal mode, the optimal gap value should be selected in the range from − 1 to 1.

Note that, with increasing viscous friction μ values of the impact velocity decreases.

When tuning the system, determining such parameter values $\lambda = \lambda^*$ at which impact velocity changes little or remains constant with changing the gap is of considerable practical interest. Determining the parameters k, m, ω, at which the value $\dot{\xi}_0$ depended little on the reduced gap ξ_0 allows one to tune the vibro-impact mechanism to a certain value of the impact velocity at any gap. Figure 4.11 presents the lines of equal impact velocities for $R = 0.6$ and $\mu = 0.1$.

The lines in given figures are seen to be plotted for parameter values satisfying the inequality $\lambda > \mu$, which in dimensionless parameters can be rewritten in the form

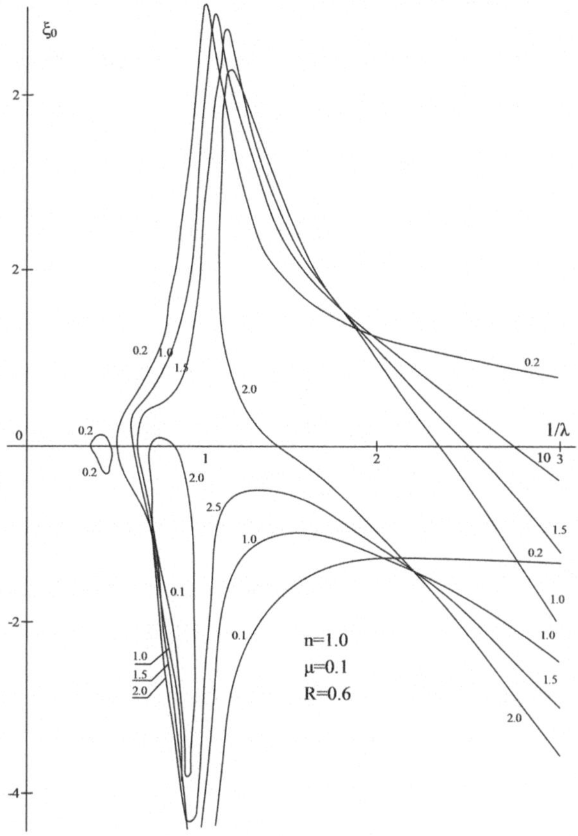

Fig. 4.11 Lines of equal impact velocities for $R = 0.6$ and $\mu = 0.1$

$$\sqrt{\frac{K}{M}} > \frac{h}{2M}$$

When this relationship holds, oscillatory motions between the parameters are seen to occur in the system.

$$h > h_k = 2M\sqrt{\frac{K}{M}}$$

where h_k is so-called critical attenuation coefficient [18]. Then the eigen motions are no oscillatory in nature but are slow motions to a certain equilibrium position. This case of high viscous friction forces has not been investigated in the present paper.

It follows from the above results that the widest part of the stability region is in the neighborhood $\lambda = 1$, i.e., for parameters satisfying the resonance condition. Therefore, this case of resonance ($\lambda = 1$) is of great interest. The motion of the system is characterized by three parameters ξ_0, R, μ. Point mapping S of the plane Π into itself has the form

$$\begin{aligned}
\xi_0 &= -\frac{\cos\tau_1}{2\mu} + \mathrm{e}^{-\mu(\tau_1-\tau_0)}\big[\alpha^{-1}A\sin\alpha(\tau_1-\tau_0) \\
&\quad + B\cos\alpha(\tau_1-\tau_0)\big] \\
\dot{\xi}_1 &= \frac{\sin\tau_1}{2\mu} + \mathrm{e}^{-\mu(\tau_1-\tau_0)}\big[(A-\mu B)\cos\alpha(\tau_1-\tau_0) \\
&\quad - \big(\alpha B + \mu\alpha^{-1}A\big)\sin\alpha(\tau_1-\tau_0)\big]
\end{aligned} \tag{4.1.29}$$

where

$$\begin{aligned}
A &= \left[\xi_0\mu - R\dot{\xi}_0 - \frac{1}{2}(\mu\sin\tau_0 - \cos\tau_0)\right] \\
B &= \big(\xi_0 + (2\mu)^{-1}\cos\tau_0\big),\ \ \alpha = \sqrt{1-\mu^2}
\end{aligned} \tag{4.1.30}$$

It follows from (4.1.30) that damping coefficient varies from 0 to 1.

The phase τ_0 is obtained from the following system of equations

$$\begin{aligned}
\sin\tau_0 &= 2\mu p^{-1}\big(g - \mathrm{e}^{-d}F\big)\dot{\xi}_0 \\
\cos\tau_0 &= -2\mu\big(\xi_0 + fp^{-1}\mathrm{e}^{-d}\dot{\xi}_0\big)
\end{aligned} \tag{4.1.31}$$

Boundary N_+, coinciding with the boundary of existence, has the form.

$$\begin{aligned}
|\xi_0| &= \zeta \quad \text{at}\, f\xi_0 > 0. \\
|\xi_0| &= \zeta\sqrt{1 + \frac{\mathrm{e}^{-2d}f^2}{(g - \mathrm{e}^{-d}F)^2}} \quad \text{at}\, f\xi_0 > 0
\end{aligned} \tag{4.1.32}$$

Boundary N_- is written in the form

$$|\xi_0| = \zeta\sqrt{\frac{(g - e^{-d}F)^2 + e^{-2d}f^2}{(g - e^{-d}F)^2 + 4e^{-2d}f^2\left(1 + R^2e^{-2d}\right)}} \tag{4.1.33}$$

Boundary C_τ is obtained from (4.1.23), (4.1.24), assuming that $\lambda = 1$.

Since the parameter space of the system is three-dimensional, a complete partition of the 3D parameter space into stability regions was carried out according to the corresponding formulas. For this, C_τ surfaces and N-stability for $n = 1$ were constructed. The region of stability of a single-impact periodic mode lies between the coordinate surfaces $\mu = 0$, $R = 0$ and surfaces N_+, N_- Fig. 4.12.

The bifurcation surface C_τ divides this region into two. A single-impact periodic motion takes place in the region lying to the right of the surface C_τ (the shaded part of the space). For the parameters in the domain to the left of this surface, a single-impact motion will first turn into two-impact and then into a more complex multi-impact.

Equations (4.1.23) and (4.1.24) determining the bifurcation surface of additional impacts showed that periodic modes with a multiplicity greater than unity do not exist, since at $\lambda = 1$, Eq. (4.1.23) has the solution only for $n = 1$. A similar result holds for a vibration striker without viscous friction [3].

Figure 4.13 presents the plotted surfaces of equal impact velocities for values 0.2, 0.6, 0.8, 1.0, 2.0 in the domain of single-impact periodic motion. Within creasing values of impact velocity, the surfaces of equal impact velocities approach the plane

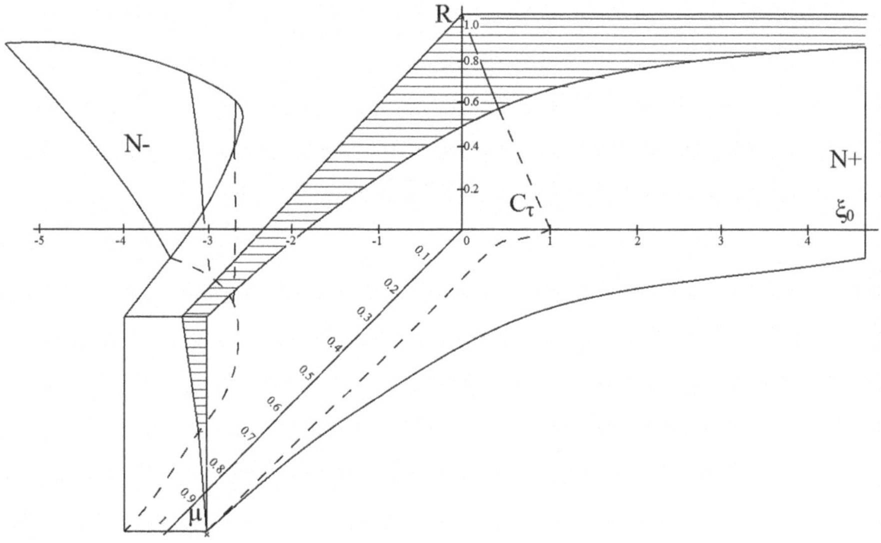

Fig. 4.12 The stability region of the single-impact periodic mode is between the coordinate surfaces $\mu = 0$, $R = 0$ and the surfaces N_+, N_-

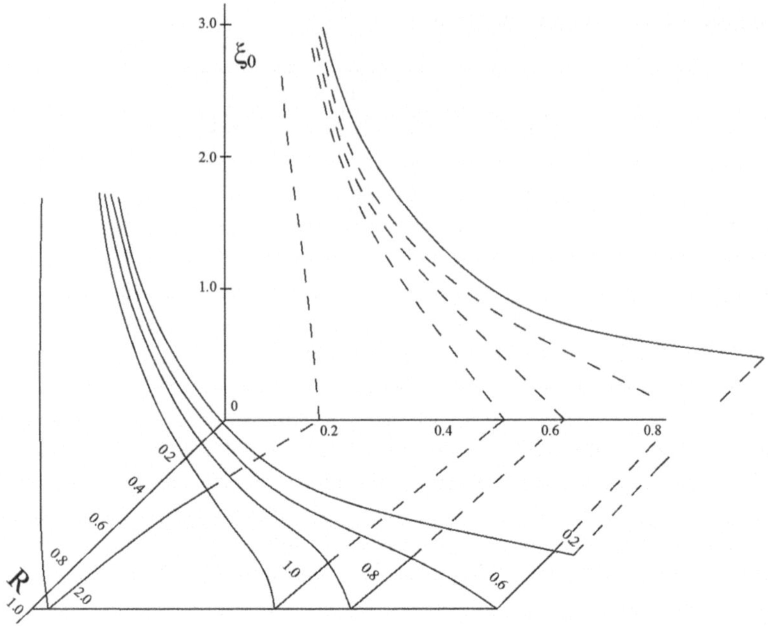

Fig. 4.13 In the region of single-impact periodic motion, surfaces of equal impact velocities are given for values equal to 0.2, 0.6, 0.8, 1.0, 2.0

$\mu = 0$. It can be seen from the behavior of these surfaces that with increasing μ, the gap should be reduced in order to maintain a constant impact velocity.

4.2 The Effect of Viscous Friction on Stable Operation of the Two-Mass Vibro-impact Mechanism

Let dissipative forces, proportional (coefficient of proportionality h_1, h_2) to the first degree of mass rate m_1, m_2, emerge during the motion of a vibro-striker. Then in dimensionless variables

$$\xi_1 = \frac{m_1\omega^2}{F}x_1 - \frac{Q_1 + Q_2}{F\lambda_1^2}, \quad \omega t = \tau$$

$$\xi_2 = \frac{m_2\omega^2}{F}x_2 - \frac{\phi\lambda_2^2 Q_1 + \left(\lambda_1^2 + \phi\lambda_2^2\right)Q_2}{\lambda_1^2\lambda_2^2 F} \tag{4.2.1}$$

Equations of system motion will have the form at $\xi_1 < \xi_0$

$$\ddot{\xi}_1 + V_1\dot{\xi}_1 + \lambda_2^2(\phi\xi_1 - \xi_2) + \lambda_1^2\xi_1 = 0$$

$$\ddot{\xi}_2 + V_2\dot{\xi}_2 + \lambda_2^2(\xi_2 - \phi\xi_1) = \sin\tau \tag{4.2.2}$$

at $\xi_1 = \xi_0, \dot{\xi}_1 = -R\dot{\xi}_0$.
where dimensionless parameters denote

$$\lambda_1^2 = \frac{c_1}{m_1\omega^2},\ \lambda_2^2 = \frac{c_2}{m_2\omega^2},\ \phi = \frac{m_2}{m_1}$$
$$V_1 = \frac{h_1}{m_1\omega},\ V_2 = \frac{h_2}{m_2\omega} \tag{4.2.3}$$

The motions defined by Eqs. (4.2.2) occur in the subspace $\xi_1 < \xi_0$ of 5D phase space $\xi_1, \dot{\xi}_1, \xi_2, \dot{\xi}_2, \tau$. To study periodic motions, we consider the point mapping of a hyperplane $\xi_1 = \xi_0$ into itself.

Let the initial image point $M_0\left(\xi_1 = \xi_0, -R\dot{\xi}_0, \xi_2^0, \dot{\xi}_2^0, \tau_0\right)$ corresponds to time immediately after the next impact of the lower mass against the stationary limiter, and the transformed point $M_1\left\{\xi_1, \dot{\xi}_1, \xi_2, \dot{\xi}_2, \tau_1\right\}$ to the time immediately before the next impact. The relationship between the coordinates of the end mapping points is obtained by solving Eqs. (4.2.2) and is written in the following form

$$\xi_0 = \sum_{i=1}^{2\sum_1 \cos\tau_1} e^{\alpha_i(\tau_1-\tau_0)}\left[C_{2i-1}^0 c_i + C_{2i}^0 \varsigma_i\right] + a\sin$$

$$\xi_2^1 = \sum_{i=1}^{2} e^{\alpha_i(\tau_1-\tau_0)}[(C_{2i-1}^0 K_{2i-1} + C_{2i}^0 K_{2i})c_i$$
$$+ \left(C_{2i}^0 K_{2i-1} - C_{2i-1}^0 K_{2i}\right)\varsigma_i] + c\sin\tau_1 + d\cos\tau_1$$
$$c_i = \cos\eta_i(\tau_1 - \tau_0),\ \varsigma_i = \sin\eta_i(\tau_1 - \tau_0)$$

$$\dot{\xi}_1^1 = \sum_{i=1}^{2} e^{\alpha_i(\tau_1-\tau_0)}\left[(C_{2i-1}^0\alpha_i + C_{2i}^0\eta_i)c_i\right.$$
$$\left.+(C_{2i}^0\alpha_i - C_{2i-1}^0\eta_i)\varsigma_i\right] + a\cos\tau_1 - b\sin\tau_1$$

$$\dot{\xi}_2^1 = \sum_{i=1}^{2} e^{\alpha_i(\tau_1-\tau_0)}[(C_{2i-1}^0 K_{2i-1}^* + C_{2i}^0 K_{2i}^*)c_i$$
$$+ \left(C_{2i}^0 K_{2i-1}^* - C_{2i-1}^0 K_{2i}^*\right)\varsigma_i] + c\cos\tau_1 - d\sin\tau_1 \tag{4.2.4}$$

where $z_{1,2} = d_1 \pm i\eta_1, z_{3,4} = \alpha_2 \pm i\eta_2$ are the roots of frequency equation

$$z^4 + A_1z^3 + A_2z^2 + A_3z + A_4 = 0 \tag{4.2.5}$$

the coefficients of which are determined by the system parameters as follows

$$A_1 = V_1 + V_2, A_2 = \left(\lambda_2^2 + 1 + V_1V_2 + e\right)$$

$$A_3 = \left[V_1\lambda_2^2 + V_2(e+1)\right], A_4 = \lambda_1^2\lambda_2^2 \tag{4.2.6}$$

Amplitudes of forced oscillations $A = \sqrt{a^2+b^2}, B = \sqrt{c^2+d^2}$ of upper and lower mass are determined, respectively

$$\begin{aligned} a &= \lambda_2^2\Delta^{-1}(1 - A_2 + A_4), b = \lambda_2^2\Delta^{-1}(A_1 - A_3) \\ c &= \lambda_2^{-2}(al - V_1 b), d = \lambda_2^{-2}(aV_1 + bl) \end{aligned} \tag{4.2.7}$$

where the following notations are introduced

$$\Delta = (A_1 - A_3)^2 + (1 - A_2 + A_4)^2, l = \left(\phi\lambda_2^2 + \lambda_1^2 - 1\right) \tag{4.2.8}$$

Distribution factors $k_i, k_i^*(i = 1, 2, 3, 4)$ have the form

$$\begin{aligned} k_{2i-1} &= \lambda_2^{-2}\left(\alpha_i^2 - \eta_i^2 + V_1\alpha_i + l + 1\right), k_{2i}^* = k_{2i}\alpha_i + k_{2i-1}\eta_i \\ k_{2i} &= \eta_i\lambda_2^{-2}(\alpha_i + V_1), k_{2i-1}^* = (k_{2i-1}\alpha_i - k_{2i}\eta_i) \end{aligned} \tag{4.2.9}$$

The arbitrary constants C_i^0 in Eqs. (4.2.4) depend on the parameters of the system as follows

$$\Delta_1 C_i^0 = \xi_0 A_{i1} + \xi_2^0 A_{i2} + R\dot{\xi}_0 A_{i3} + \dot{\xi}_2^0 A_{i4} + N_i \cos\tau_0 - M_i \sin\tau_0 \tag{4.2.10}$$

where

$$\begin{aligned} A_{11} &= \left(\alpha_2 A_{13} - k_3 A_{12} - k_3^* A_{14}\right), \\ A_{12} &= \left(k_4^*\eta_1 - k_2^*\eta_2\right) = -A_{32} \\ A_{13} &= -A_{33} = \left(k_2 k_4^* - k_2^* k_4\right), \\ A_{14} &= -A_{34} = (k_2\eta_2 - k_4\eta_1) \\ A_{2i,1} &= \left(\alpha_i A_{2i,3} - k_1 A_{2i,2} - k_1^* A_{2i,4}\right), \\ A_{22} &= \left[(\alpha_2 - \alpha_1)k_4^* - \eta_2\left(k_3^* - k_1^*\right)\right], \\ A_{23} &= \left[(k_3 - k_1)k_4^* - \left(k_3^* - k_1^*\right)k_4\right], \\ A_{24} &= [(k_3 - k_1)\eta_2 - k_4(\alpha_2 - \alpha_1)], \\ A_{42} &= \left[\left(k_3^* - k_1^*\right)\eta_1 - k_2^*(\alpha_2 - \alpha_1)\right], \\ A_{43} &= \left[\left(k_3^* - k_1^*\right)k_2 - (k_3 - k_1)k_2^*\right] \\ A_{44} &= [k_2(\alpha_2 - \alpha_1) - (k_3 - k_1)\eta_1, \\ \Delta_1 &= [(\alpha_2 - \alpha_1)A_{13} - \eta_1 A_{23} - \eta_2 A_{43}] \\ M_i &= (A_{i1}a + cA_{i2} + bA_{i3} - dA_{i4}) \\ N_i &= (A_{i3}a - bA_{i1} - A_{i4}c - A_{i2}d) \end{aligned} \tag{4.2.11}$$

The coordinates of the fixed point of the point mapping corresponding to a single-impact n-fold periodic motion are derived from (4.2.2.4) by supplementing them with the conditions

$$\dot{\xi}_1^1 = \dot{\xi}_0, \xi_2^1 = \xi_2^0, \dot{\xi}_2^1 = \dot{\xi}_2^0, \tau_1 - \tau_0 = T \tag{4.2.12}$$

In order to distinguish in the parameter space $\lambda_1, \phi, \zeta, V_1, V_2, \xi_0, R$ the existence and stability regions of the considered periodic regime, we study the stability of fixed mapping points.

The coordinates of the fixed point and the characteristic polynomial (to determine the stability of periodic motion) are obtained with the help of computer simulation.

Figures 4.14, 4.15, 4.16 and 4.17 present in plane ξ_0, λ the stability regions of single-impact periodic motions for fixed values $\phi = \frac{m_2}{m_1}, \zeta = C_1/C_2, \chi = h_1/h_2$, equal to 1, $R = 0{,}1$ and for various values of viscous friction coefficients equal to 0.1; 0.2; 0.3; 0.5, respectively. With increasing V, the stability regions, as in the case of the single-mass model, become smoother due to smoothing of infinite discontinuities of the boundaries of the stability regions. The region dimensions are getting significantly narrower with increasing V over the gap. Thus, at $V = 0$, the stability regions are located over ξ_0 from $-\infty$ to $+\infty$, whereas, at $V = 0.5$, they are in the range from $-$ 1 to 1.

We study the effect of the damping coefficients χ on the value and location of stability regions. It turned out that at $\chi > 1$ (the viscosity coefficient for the lower mass is greater than that for upper), the stability regions are much larger and more indented than at $\chi < 1$.This can be explained by the fact that at $\chi > 1$ the driving force $F \sin \omega t$ can impose a periodic motion with a period multiple of its period in most of the parameter space. The situation is different when the lower damper is small and the upper one is large. In this case, the upper damper prevents the vibrator

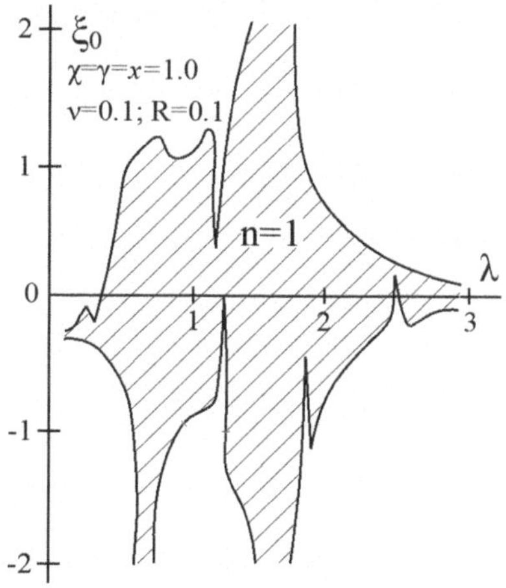

Fig. 4.14 Stability regions of single-impact single periodic movements for fixed values $\phi = m_2/m_1, \zeta = C_1/C_2, \chi = h_1/h_2$ equal to 1, $R = 0.1$ and for the viscous friction coefficient $\nu = 0.1$

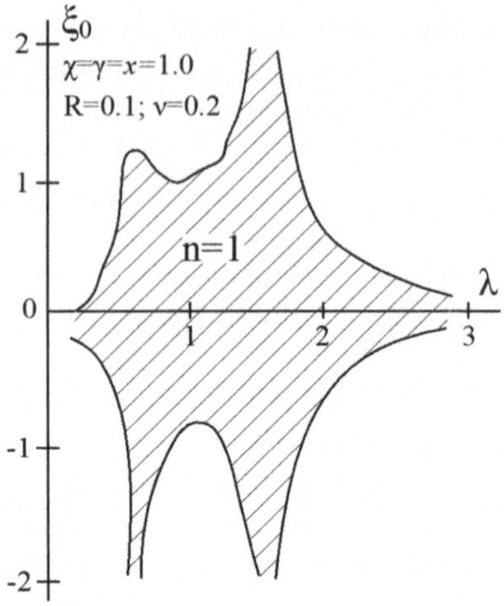

Fig. 4.15 Stability regions of single-impact single periodic movements for fixed values $\phi = m_2/m_1$, $\zeta = C_1/C_2$, $\chi = h_1/h_2$ equal to 1, $R = 0.1$ and for the viscous friction coefficient $\nu = 0.2$

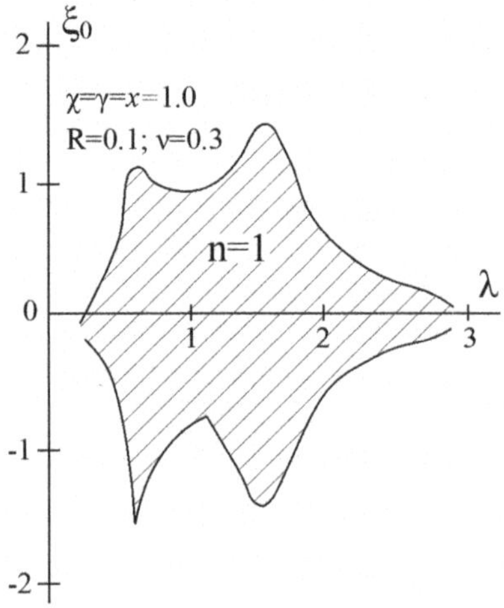

Fig. 4.16 Stability regions of single-impact single periodic movements for fixed values $\phi = m_2/m_1$, $\zeta = C_1/C_2$, $\chi = h_1/h_2$ equal to 1, $R = 0.1$ and for the viscous friction coefficient $\nu = 0.3$

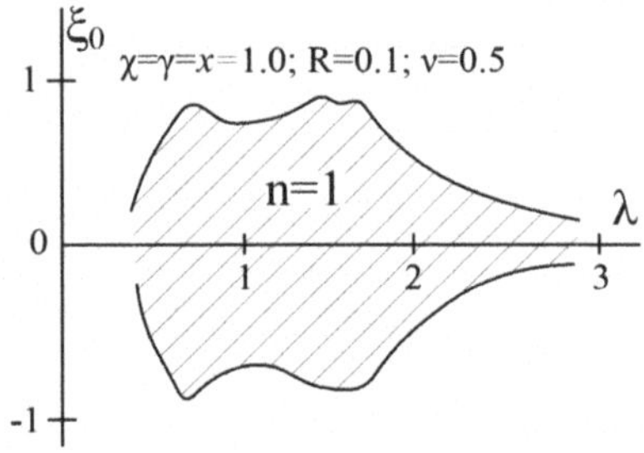

Fig. 4.17 Stability regions of single-impact single periodic movements for fixed values $\phi = m_2/m_1$, $\zeta = C_1/C_2$, $\chi = h_1/h_2$ equal to 1, $R = 0.1$ and for the viscous friction coefficient $\nu = 0.5$

from imposing periodic motions in most of the plane, but the regions themselves become smoother.

Figures 4.18 and 4.19 present the stability regions of one-impact periodic motions for $\chi = 4.0$ and $\chi = 0.25$, respectively. Studies have shown that with increasing ϕ, ζ, stability regions shift toward large λ_1 and their boundaries become smooth, similar to an increase in V. Thus, Figs. 4.20 and 4.21 show the stability regions for $\phi = 3, \zeta = 3$, respectively.

The impact velocity $\dot{\xi}_0$ was studied in all stability regions. Figures 4.22, 4.23, 4.24 and 4.25 show gap dependences if the maximum impact velocity for different parameters. It follows that the magnitude of maximum impact velocity $\dot{\xi}_0^{max}$ decreases with increasing V (with increasing the lower damping coefficient h_1). With increasing R, the magnitude of impact velocity increases (see Figs. 4.22, 4.23, 4.24 and 4.25) but in this case, the stability regions become more indented. With increasing $\chi \dot{\xi}_0^{max}$,

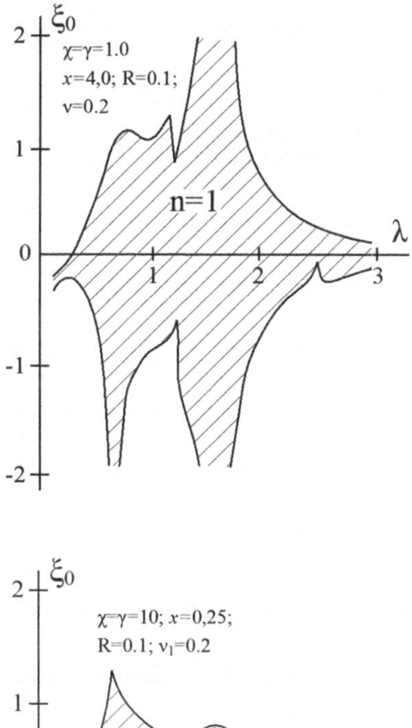

Fig. 4.18 Areas of stability of single-impact periodic movements for $\chi = 4.0$

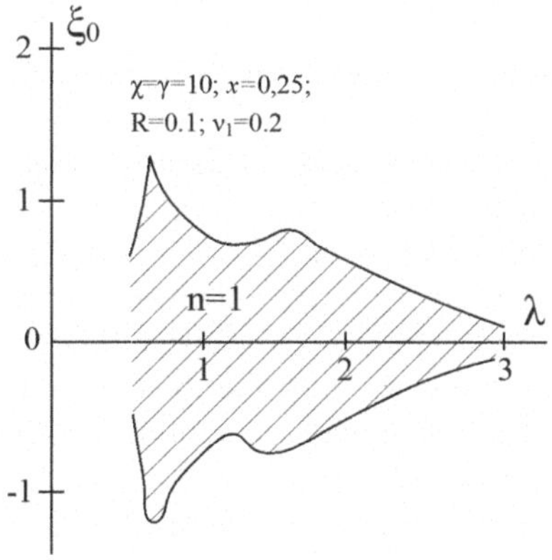

Fig. 4.19 Areas of stability of single-impact periodic movements for $x = 0.25$

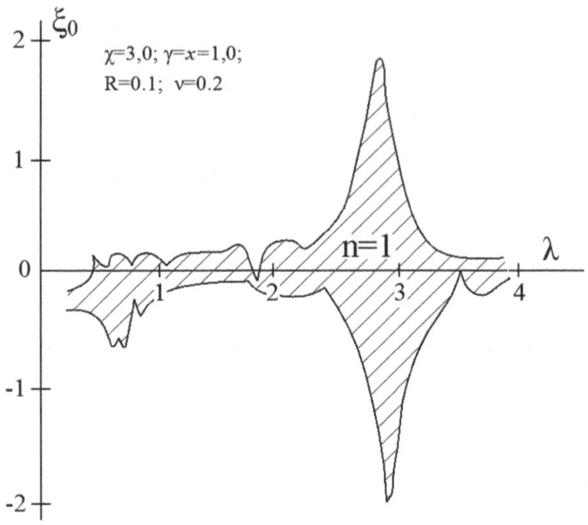

Fig. 4.20 Areas of stability of single-impact periodic movements for $\chi = 3$

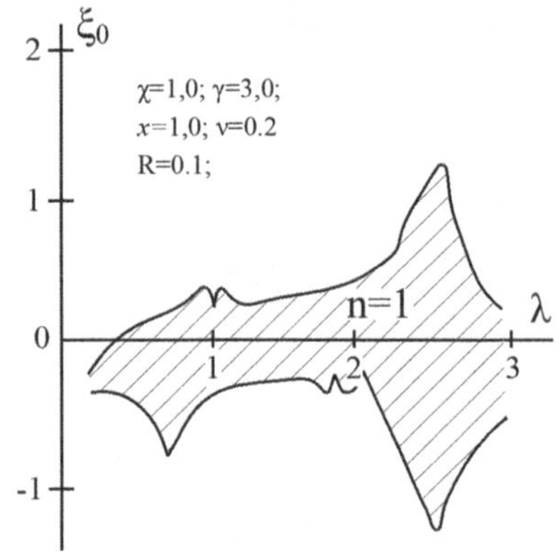

Fig. 4.21 Areas of stability of single-impact periodic movements for $\gamma = 3$

impact velocity increases, and with increasing ϕ, ζ, the value $\dot{\xi}_0^{\max}$ decreases (see Fig. 4.24).

The magnitude of the maximum impact velocity at fixed gaps as a function of λ has maxima series, decreasing when λ tends to zero and to infinity. Figure 4.25, Fig. 4.26 present $\dot{\xi}_0^{\max}$ as a function of λ at fixed $\phi = \zeta = \chi = 1$ and varying V, respectively, for $R = 0.6$ and $R = 0.1$.

It can be seen from above figures that the maximum values of impact velocity are located in the neighborhood of the resonance frequencies of the corresponding two-mass system without a limiter. With increasing V, maximum value $\dot{\xi}_0^{\max}$ is observed in the neighborhood of zero gap.

From the results obtained in § 4.1 and 4.2 of this chapter, we may conclude the following:

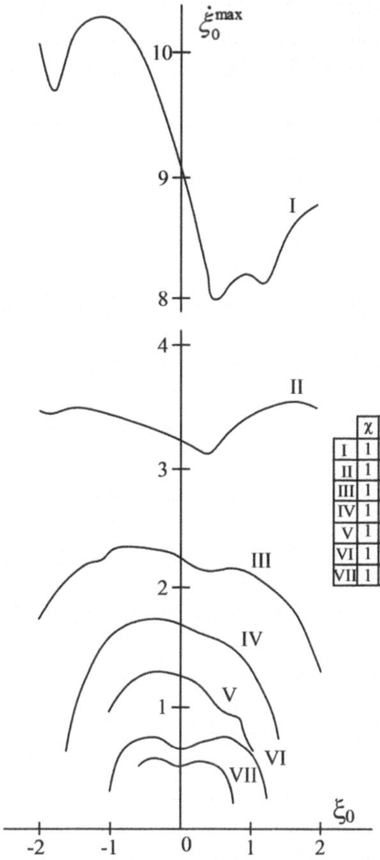

	χ	γ	x	R	ν_1
I	1	1	1	0.6	0
II	1	1	1	0.6	0.1
III	1	1	1	0.6	0.2
IV	1	1	1	0.6	0.3
V	1	1	1	0.6	0.4
VI	1	3	1	0.6	0.2
VII	1	3	1	0.6	0.3

Fig. 4.22 Dependences of the maximum impact speed on the gap for a different set of parameters

1. The stability regions of periodic motions with account for viscous friction forces, become smoother due to joining of separate parts of stability regions. However, it should be kept in view that the stability regions decrease (especially with respect to the gap). As a result, stability of single-impact motion is possible in a limited part of the plane. Thus, for $V > 0.5$, stability regions are located for the gap values satisfying the inequality $|\xi_0| < 1$.

Note that for these gaps, the maximum impact velocity lies within the stability region, which greatly simplifies the system tuning to the optimal mode, the condition when impact velocity reaches its maximum is chosen as the optimality criterion.

2. Within creasing damping coefficients h_1, h_2, the value of impact velocity decreases and the largest values of the maximum impact velocity are grouped in the neighborhood of zero gap. Hence, it directly follows that in order to tune the system to the optimal mode, the reduced gap must be chosen in the neighborhood of zero gap.

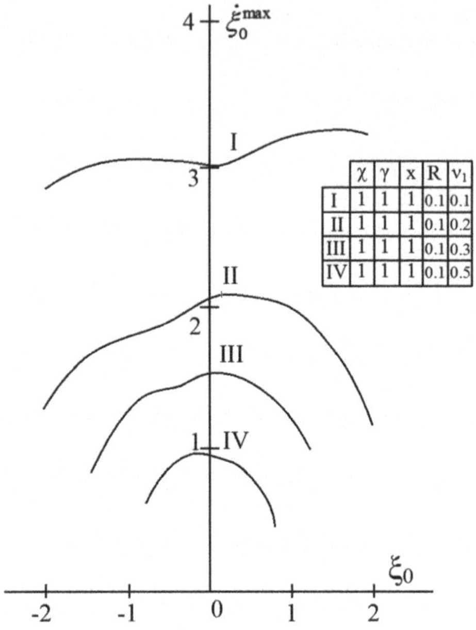

Fig. 4.23 Dependences of the maximum impact speed on the gap for a different set of parameters

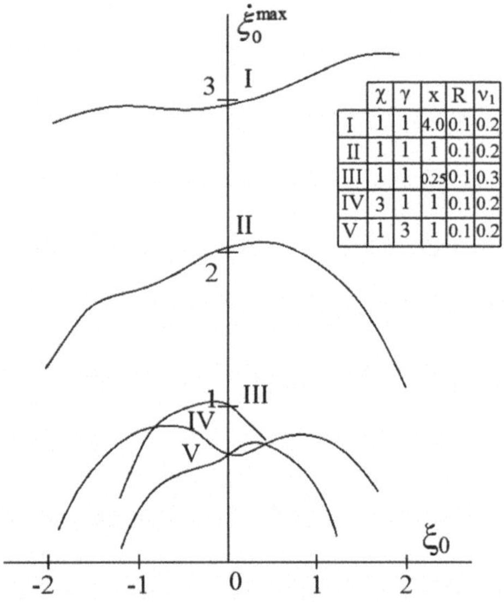

Fig. 4.24 Dependences of the maximum impact speed on the gap for a different set of parameters

3. If the viscous friction forces are large for the lower mass and small for the upper one ($\chi > 1$), then the stability regions expand significantly and vice versa, if $\chi < 1$, then the stability regions become narrower with respect to ξ_0.

Fig. 4.25 Dependences of the maximum impact speed on the gap for a different set of parameters

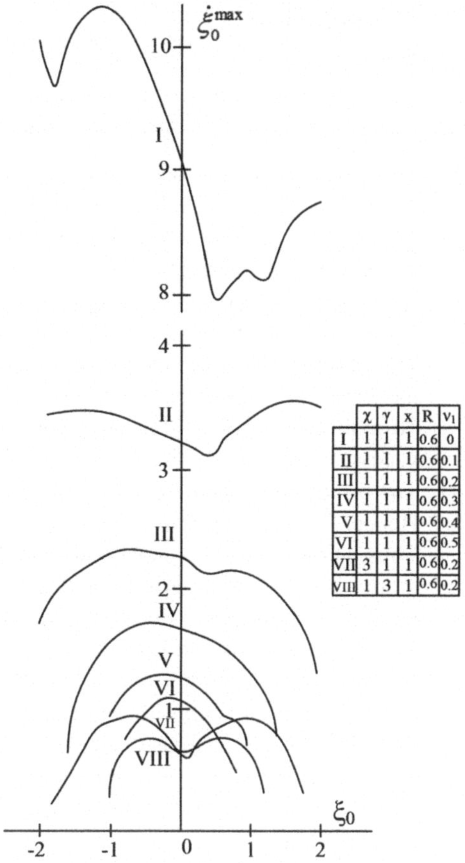

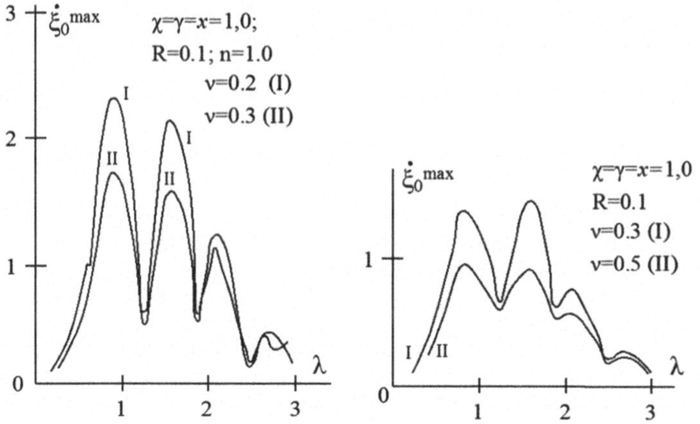

Fig. 4.26 Dependences of the maximum impact velocity λ on for a different set of parameters

4.3 Conclusion

In both single-mass and two-mass vibro-impact mechanisms, the effect of viscous friction forces on stability of periodic motions is reduced to the following;

(1) The scattered parts of the stability region (at $\mu = 0$) of single-impact n-fold are connected (for $\mu \neq 0$) due to smoothing of infinite boundary discontinuities.
(2) The stability regions are not only smoothed, but also significantly narrowed, especially with respect to the gap ξ_0. Thus, for $\mu = 0.5$, they lie with in $|\xi_0| < 1$.Therefore, with a large energy dispersion between impacts, maintaining a stable periodic regime is possible in a rather narrow range of parameter values. This emphasizes once again the need for theoretical study of vibro-impact mechanisms in order to tune them to the required periodic mode.
(3) To maintain periodic motion taking into account viscous friction forces, it is necessary to reduce the driving force frequency ω.
(4) The bifurcation surface C_τ cuts off a significant part from the stability region. Hence, constructing this surface in the parameter space is of great importance. Thus, in the case of resonance in a single-mass mechanism, there exist periodic motions only with a period equal to that of the external force. Motions with a large period disappear due to the fact that all the parameter values R, ξ_0, μ (at $\lambda = 1$) are discontinuity points of point transformation.
(5) In some plane sections $1/\lambda, \xi_0$ of 4D parameter space, the superposition of periodic motions with different multiplicities is observed to disappear.
(6) Stability regions decrease with increasing viscous friction forces for positive gaps and increase for negative gaps. It immediately follows that the gap must be chosen in the neighborhood of zero or negative values (with preliminary compression).
(7) The magnitude of impact velocities decreases with increasing damping coefficient and reaches its maximum (in the case of a single-mass mechanism) for periodic motion with a period of external force.

References

1. Brogliato, B.: Some perspectives on the analysis and control of complementarity systems. IEEE Trans. Autom. Control **48**(6), 918–935 (2003)
2. Bartolo, D., Boudaoud, A., Narcy, G., Bonn, D.: Dynamics of non-Newtonian droplets. Phys. Rev. Lett. **99**, 174502 (2007)
3. Bespalova, L.V.: To the theory of vibro-impact mechanism. Izvestia of the Academy of Sciences of the USSR, OTN, No. 5 (1957)
4. Neimark, Yu.I.: The method of point transformations in the theory of nonlinear oscillations. Izv. higher uch. Head, Radiophysics **1** (1958)
5. Neimark Yu.I.: The method of point mappings in the theory of nonlinear oscillations. Izv. Vysh. Uch. Head., Radiophys. **2**(2) (1958)
6. Neimark, Yu.I.: Method of point transformations in the theory of nonlinear oscillations. Izv. Higher Uch. Head, Radiophys. **3**(5–6) (1958)

7. Bespalova, L.V., Neimark, Yu.I., Feigin, M.I.: Dynamical systems with shock interactions and the theory of nonlinear oscillations. Engineering MTT Magazine, No. 1 (1966)
8. Oldenburg, R., Sartorius, G.: Dynamics of automatic control. GEI M-L (1949)
9. Kobrinsky, A.E.: Mechanisms with Elastic Links, 390 p. Nauka (1964)
10. Feigin, M.I.: On the theory of nonlinear dampers. Izv. Higher Uch. Head, Radiophys. **2**(4) (1959)
11. Boyarinov, V.S., Neimark, Yu.I.: On shaft vibrations in a ball bearing. Izv. USSR Acad. Sci. Mech. **3** (1965)
12. Podchufarov, B.M.: Dynamics of cyclic automation of a machine with a movable base. Izv. Vys. Uch. Zav., Mashinostroyeniye **6** (1963)
13. Bespalova, L.V.: The results of simulation of the problem of vibration and vibroimpact immersion. Izv. Higher Educ. Manag. Radiophys. **3**(1) (1960)
14. Bespalova, L.V.: On the theory of vibroimpact driving. Izv. Vys. Acad. Radiophys. **2**(4) (1959)
15. Goldin, Yu.M.: Comparative study of one-mass and two-mass vibroimpact mechanisms. On Sat. "Research of construction and road machines with the use of computer technology". TsNIITE stroymash (1968)
16. Batalova 3.S. Study of vibration penetration of a sheet pile with frontal soil resistance. Engineering Magazine, MTT, No. 1 (1966)
17. Silvestrov, E.E., Shlyakhtin, A.V.: Movement of a vibrational system under the action of a time-dependent force. In: Proceedings of the seminar on TMM, issue 85–86 (1961)
18. Galeani, S., Menini, L., Potini, A., Tornambй, A.: Trajectory tracking for a particle in elliptical billiards. Int. J. Control. **81**(2), 189–213 (2008)

Chapter 5
Dynamics of Vibro-impact Systems with Randomly Changing Parameters

Abstract Chapter five is devoted to the study of the nonlinear dynamics of single-mass vibration-impact systems with small random changes in the recovery coefficient of speed and gap from those values at which a stable periodic regime exists in the deterministic case. For the first time, statistical characteristics have been determined in cases of delta-correlated and weekta-correlated changes in the coefficient of restoration and clearance; the values of parameters have been found at which small random deviations of the coefficient of restoration do not affect the single-impact mode of movement, as well as those parameters at which week-correlated changes have the same effect on the operation of the vibration-impact mechanism as with delta-correlated changes in the recovery coefficient of speed and clearance. A new concept of stochastic stability is introduced based on existing definitions. Based on the introduced definition, one can judge the degree of stochastic stability of the movement under consideration. Based on this, some practical recommendations have been made for selecting system parameters. The influence of random changes in the speed recovery coefficient on the value of the optimal impact speed has been studied. Digital modeling of the problem of the movement of a vibratory impactor with a random change in parameters has been carried out. The simulation results agree well with the theoretical ones with parameter deviations ranging from -0.05 to 0.05. An approximate value for the smallness of random changes has been found.

Keywords Point mapping method · Delta and not delta correlated random variables · Markov process · Stochastic stability · Difference equations

During the operation of mechanisms with impact interaction, used in numerous branches of technology, many parameters, such as dynamic characteristics of the processed medium and, naturally, the associated coefficient of restitution upon impact, gaps, weight characteristics, and many others, do not remain constant, but change in the process of prolonged work, randomly from impact to impact. Hence, a complete study of vibro-impact mechanisms leads to studying the dynamics of the systems with randomly changing parameters.

V. Metrikin and I. Leonid, *Forced Oscillations of Multidimensional Highly Nonlinear Systems*, Advanced Structured Materials 222,
https://doi.org/10.1007/978-981-97-6779-3_5

Some studies of the dynamics of mainly linear systems with random perturbations were carried out by V. I. Babitsky, Yu. M. Goldin, L. B. Zaretsky, D. A. Kobriksky, M. Z. Kolovsky, S. K. Lukomsky, and others.

The point mapping method can also be employed for studying the dynamics of nonlinear systems with random effects. This method was first applied in the work of S. M. Rytov for research of self-oscillating systems with random perturbations. In this work, fluctuations of piecewise linear systems with one degree of freedom, caused by small and short-correlated random forces, were investigated. Further, R. X. Sadekova generalizes Rytov's method for the case of non-delta-correlated noise affecting a piecewise linear system with an arbitrary number of degrees of freedom.

Below, the method of point mappings is applied to studying the dynamics of vibro-impact systems with randomly changing parameters.

In the first section, fluctuations б of single-mass vibro-impact mechanisms are studied theoretically for small random deviations of the coefficient of restitution and the gap from those values for which a stable periodic mode exists in the deterministic case. Random parameter deviations can be either delta-correlated or non-delta-correlated. In this section, the correlation effect on the statistical system characteristics is investigated.

In the second section, the concept of stochastic stability is introduced based on the results obtained by V. A. Brusin, N. A. Katz, N. N. Krasovsky, Sh. O. Nevelson, M. L. Tyem, R. Z. Khasminsky, and others. The introduced concept of stochastic stability made it possible to determine the optimal values of the system parameters.

In the third section, we study the effect of random deviations of the coefficient of restitution by the value of optimal impact pulse. The parameter space is divided into two subspaces, in one of which the value of the optimal impact velocity increases, and in the other, it decreases. At the boundary of these two subspaces, the magnitude of the maximum impact velocity does not change.

In the fourth section, the probability of P_B system exit from a specified mode is theoretically studied. The value P_B is evaluated above as a function of the size of domain attraction of the fixed point of point mapping of the deterministic process and of the magnitude of random deviations R.

In the fifth section, theoretical results obtained in Chaps. 1 to 4 are verified, and the magnitude of "smallness" random deviations R is estimated. It turned out that under small deviations we can mean the variation value R in the range of 0.05.

5.1 Investigation of Single-Mass Vibro-impact Mechanisms with Randomly Changing Parameters

Let us consider a single-mass vibro-striker, which consists of a mass M that is elastically fixed with a coefficient of elasticity k, subject to external periodic force $F \sin \omega t$ (Fig. 5.1). Under the assumptions A/see Sect. 5.1/, differential equations of the system in the interval between impacts are written in the form ($x < x_0$)

$$M\ddot{x} + kx = \Phi + F \sin \omega t \quad (5.1.1)$$

where Φ is constant force acting upon impact mass, and $\Phi \neq mg$, F, ω are the amplitude and frequency of the driving force.

When $x = x_0$, an inelastic impact with a coefficient of restitution R occurs, so that if $\dot{x}_0^-$ is striker velocity immediately before impact, then $\dot{x}_0^+$ is the velocity after the impact and $\dot{x}_0^+ = -R\dot{x}_0^-$.

In dimensionless variables

$$\xi = \frac{M\omega^2}{F}\left(x - \Phi\kappa^{-1}\right) \quad (5.1.2)$$

Equation (5.1.1) will be rewritten

$$\ddot{\xi} + \lambda^2\xi = \sin \tau, \text{ at } \xi < \xi_0 \quad (5.1.3)$$

$$\dot{\xi}_0^+ = -R\dot{\xi}_0^-, \text{ at } \xi = \xi_0 \quad (5.1.4)$$

The phase space and the classification of periodic motions of the considered system are studied in detail in the work of Bespalova [1]. Using the results of this work, the point mapping can be written as follows:

$$\begin{aligned}
\xi_0 &= \xi_0 \cos \lambda(\tau_n - \tau_{n-1}) - \alpha \sin \tau_{n-1} \cos(\tau_n - \tau_{n-1})\lambda \\
&\quad - \alpha\lambda^{-1} \cos \tau_{n-1} \sin \lambda(\tau_n - \tau_{n-1}) + \alpha \sin \tau_n \\
&\quad - R\lambda^{-1}\dot{\xi}_{n-1} \sin \lambda(\tau_n - \tau_{n-1}), \\
\dot{\xi}_n &= -\xi_0\lambda \sin \lambda(\tau_n - \tau_{n-1}) + \alpha\lambda \sin \lambda(\tau_n - \tau_{n-1}) \\
&\quad - \alpha \cos \tau_{n-1} \cos(\tau_n - \tau_{n-1})\lambda + \alpha \cos \tau_n \\
&\quad - R\dot{\xi}_{n-1} \cos \lambda(\tau_n - \tau_{n-1})
\end{aligned} \quad (5.1.5)$$

Here $\alpha = (\lambda^2 - 1)^{-1}$.

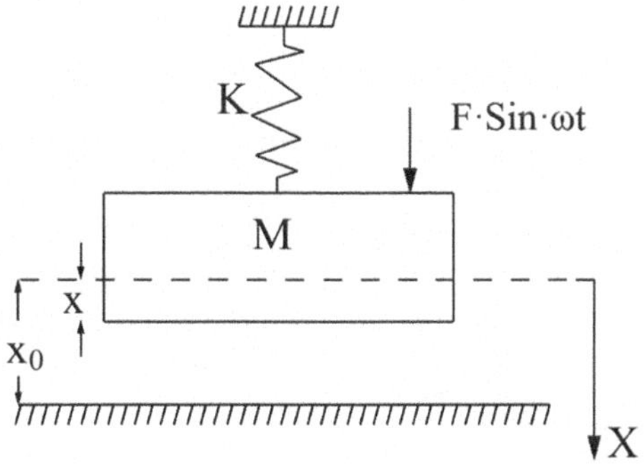

Fig. 5.1 Physical diagram of a single-mass vibratory impactor

Suppose further that during the operation of a single-mass vibro-striker the coefficient of restitution and the gap from impact to impact change randomly, and their changes are so small that the system remains in the neighborhood of a stable periodic mode.

Suppose that there exist a stable periodic motion in a deterministic system for some parameter values $R = R_0$ and $\xi = \xi_0$. The coefficient of restitution and the gap at n—impact takes the values:

$$R_n = R_0 + \triangle R_n, \xi_n = \xi_0 + \triangle \xi_n \tag{5.1.6}$$

where ΔR_n and $\Delta\xi_n$ are small random deviations of R_0 and ξ_0 with a zero mean value and a nonzero dispersion.

Point mapping (5.1.5) for random discrete / from impact to impact / deviation of parameters R and ξ_0 will be rewritten in the form

$$\begin{aligned} \xi_{n+1} &= \xi_n \cos\lambda(\tau_{n+1} - \tau_n) - \alpha \sin\tau_n \cos\lambda(\tau_{n+1} - \tau_n) \\ &\quad + \alpha \sin\tau_{n+1} - \alpha\lambda^{-1} \cos\tau_n \sin\lambda(\tau_{n+1} - \tau_n) - y_n\lambda^{-1}(\tau_{n+1} - \tau_n), \\ y_{n+1} &= R_{n+1} \left\{ \begin{array}{l} \xi_n\lambda \sin\lambda(\tau_{n+1} - \tau_n) - \alpha\lambda \sin\tau_n \sin\lambda(\tau_{n+1} - \tau_n) \\ -\alpha \cos\tau_{n+1} + \alpha \cos\tau_n \cos\lambda(\tau_{n+1} - \tau_n) \\ -y_n \cos\lambda(\tau_{n+1} - \tau_n) \end{array} \right\} \end{aligned} \tag{5.1.7}$$

where y_n denotes the velocity of the striker immediately after n—impact.

For further research, we study the properties of the system for two special cases:

1. suppose that in operation of a vibro-impact mechanism, the gap remains constant while the coefficient of restitution changes randomly;
2. the coefficient of restitution upon impact remains constant during operation of the mechanism under study for any impact, and the gap changes randomly.

Case 1. (Small random deviations of R at constant gap ξ_0).

The general point mapping (5.1.7) will be rewritten as follows:

$$\begin{aligned} \xi_0 &= \xi_0 \cos\lambda(\tau_{n+1} - \tau_n) - \alpha \sin\tau_n \cos\lambda(\tau_{n+1} - \tau_n) + \alpha \sin\tau_{n+1} \\ &\quad - \alpha\lambda^{-1} \cos\tau_n \sin\lambda(\tau_{n+1} - \tau_n) - y_n\lambda^{-1} \sin\lambda(\tau_{n+1} - \tau_n) \\ y_{n+1} &= R_{n+1} \left\{ \begin{array}{l} \xi_0\lambda \sin\lambda(\tau_{n+1} - \tau_n) - \alpha\lambda \sin\tau_n \sin\lambda(\tau_{n+1} - \tau_n) \\ -\alpha \cos\tau_{n+1} + \alpha \cos\tau_n \cos\lambda(\tau_{n+1} - \tau_n) \\ -y_n \cos\lambda(\tau_{n+1} - \tau_n) \end{array} \right\} \end{aligned} \tag{5.1.8}$$

From the point mapping, it is directly seen that random sequences of points y_1, τ_1; $y_2, \tau_2; \ldots y_{n-1}, \tau_{n-1}; \ldots$ determined from relations (5.1.8) for $R = R_1, R_2, \ldots, R_{n-1}, \ldots$, respectively, represent a Markov process to which the theory of Markov processes can be applied.

Since, under assumption that random deviations of the coefficient of restitution ΔR_n at each n—impact from R_0 are small, then set of points $\{y_\kappa, \tau_\kappa\}$ will be located in

ε neighborhood of a stable periodic mode corresponding to the value R_0. Therefore, it is possible to linearize the point mapping (5.1.8) in the neighborhood of this stable periodic motion, where after we obtain a system of equations in finite differences with respect to deviations of the after impact velocity, impact time, and coefficient of restitution in the form:

$$\begin{aligned} \Delta y_{n+1} &= (\Phi_1\Phi_2 + Q_2)\Delta\tau_n + (M + \Phi_2 Q_1)\Delta y_n - \dot{\xi}_0 \Delta R_{n+1} \\ \Delta\tau_{n+1} &= \Phi_1\Delta\tau_n + Q_1\Delta y_n \end{aligned} \tag{5.1.9}$$

where the following notations are introduced:

$$\begin{aligned} \Phi_1 &= \{1 - \alpha\nu[(1-\gamma)\cos\tau_0 - \upsilon\sin\tau_0]\},\ Q_1 = -\upsilon\nu \\ M &= -R_0\gamma,\ \Phi_2 = R_0 f_1,\ T = 2\pi n,\ \gamma = \cos\lambda T \\ \nu &= \dot{\xi}_0^{-1},\ Q_2 = -R_0\left[2^{-1}f_2 + \alpha\lambda^2\upsilon\cos\tau_0\right] \\ \upsilon &= \lambda^{-1}\sin\lambda T,\ f_1 = \sin\tau_0 - \lambda^2\xi_0 \\ f_2 &= 2\left[\lambda^2\xi_0 - \alpha\left(\lambda^2 - \gamma\right)\sin\tau_0\right] \end{aligned}$$

Further we need to solve inhomogeneous second-order difference equations, to which the system of two difference equations of the form (5.1.9) can be reduced. Suppose that after simple transformations of the system (5.1.9), we obtain a difference equation of the form:

$$\Delta\tau_{n+2} + \alpha\Delta\tau_{n+1} + \beta\Delta\tau_n = \gamma J_n \tag{5.1.10}$$

The solution to the difference equation will be sought [2] as the sum of two solutions: the general solution of the homogeneous equation, which can be written:

$$\Delta\tau_n^0 = C_1 z_1^n + C_2 z_2^n$$

where z_1, z_2 are the roots of characteristic equation

$$z^2 + \alpha z + \beta = 0$$

and private solution Δz_n^* of heterogeneous equation.

To find a particular solution, we use the method of variation of an arbitrary constant. We assume that arbitrary constants C_1 and C_2 depend on n, and then, after simple transformations, we obtain a system of equations for determining arbitrary constants in the form:

$$\begin{aligned} z_1^{n+2}\Delta C_1(n) + z_2^{n+2}\Delta C_2(n) &= \gamma J_n \\ z_1^{n+1}\Delta C_1(n) + z_2^{n+1}\Delta C_2(n) &= 0 \end{aligned}$$

Solving the resulting system with respect to $\Delta C_1(n)$ and $\Delta C_2(n)$, we obtain:

$$\Delta C_1(n) = \frac{\gamma J_n}{(z_1 - z_2)z_1^{n+1}}, \ \Delta C_2(n) = \frac{\gamma J_n}{(z_2 - z_1)z_2^{n+1}}$$

Reducing these equalities, we can obtain expressions for arbitrary constants [2], $C_1(n) = \sum_{v=0}^{n-1} \frac{\gamma J_v}{(z_1 - z_2)z_1^{v+1}}, \ C_2(n) = \sum_{v=0}^{n-1} \frac{\gamma J_v}{(z_2 - z_1)z_2^{v+1}}$.

After this, the solution of inhomogeneous difference equation can be written in the form:

$$\Delta z_n = C_1 z_1^n + C_2 z_2^n + \gamma \sum_{v=0}^{n-1} J_v \frac{z_1^{n-v-1} - z_2^{n-v-1}}{z_1 - z_2}$$

From this general solution, it follows [3] that

$$\Delta \tau_n = A_1 z_1^{n-\kappa} + A_2 z_2^{n-\kappa} + \upsilon \sum_{i=\kappa}^{n-1} \alpha_{n,i} \Delta R_{i+1}$$

$$\Delta y_n = A_1' z_1^{n-\kappa} + A_2' z_2^{n-\kappa} + \dot{\xi}_0 \sum_{i=\kappa}^{n-1} \alpha_{n,i} (\Phi_1 \Delta R_{i+1} - \Delta R_{i+2}) \tag{5.1.11}$$

where

$$\alpha_{n,i} = \frac{z_1^{n-i-1} - z_2^{n-i-1}}{z_1 - z_2} \tag{5.1.12}$$

And A_1, A_2, A_1', A_2' are arbitrary constants depending on initial values.

Assuming that the periodic motion is stable, the roots of the characteristic equation of system (5.1.8)

$$\begin{vmatrix} -z + M + Q_1 \Phi_1 & \Phi_1 \Phi_2 + Q_2 \\ Q_1 & -z + \Phi_1 \end{vmatrix} = 0 \tag{5.1.13}$$

satisfy the condition $|z_i| < 1, i = 1, 2$.

Expanding (5.1.13), we obtain a second-degree polynomial:

$$z^2 + \alpha_1 z + \beta_1 = 0 \tag{5.1.14}$$

where α_1 and β_1 coincide up to notations with the corresponding expressions in [1].

By tending the index to minus infinity in relations (5.1.10), the terms depending on the initial conditions vanish, and there remain series the coefficients of which depend only on the system parameters in the form

$$\Delta \tau_n = \upsilon \sum_{i=\kappa}^{n-1} \alpha_{n,i} \Delta R_{i+1}$$

$$\Delta y_n = \dot{\xi}_0 \sum_{i=\kappa}^{n-1} \alpha_{n,i}(\Phi_1 \Delta R_{i+1} - \Delta R_{i+2}) \tag{5.1.15}$$

From (5.1.15), one can directly derive parameter values at which random deviations of the coefficient of restitution would not affect the instant of impact. These will be the values of parameters at which the value $\upsilon = \lambda^{-1} \sin \lambda T$ vanishes, i.e.,

$$\lambda_\kappa T = \pi \kappa \tag{5.1.16}$$

However, the equality (5.1.16) includes some values λ_j, which are the boundary of the stability region N_+, in which

$$|z_i(\lambda_j)| = 1,\ \lambda_j = \frac{j}{n},\ (j \neq n, j, n = 1, 2, ...)$$

Thus, relations (5.1.15) lose their meaning. Therefore, eliminating from (5.1.16) the values corresponding to the boundaries of the stability region, we obtain that for

$$\lambda_\kappa = \frac{2\kappa + 1}{2n} \tag{5.1.17}$$

where n is the multiplicity of the period of motion to the period of external force and $\kappa = 0, 1, 2, \ldots$, and instants of impacts τ_n will be independent of random deviations R.

It is shown in the work (5.1.15) that the after impact velocities are independent of random changes of R only for those parameters at which the impact velocity, $\dot{\xi}_0$, defined by equation

$$\left[(1-R)^2\lambda^2 + (1+R)^2 ctg^2 \pi n\lambda\right]\dot{\xi}_0^2 + 4(1+R)\lambda\xi_0 ctg\pi n\lambda\dot{\xi}_0 + 4\lambda^2\left(\xi_0^2 - \alpha^2\right) = 0 \tag{5.1.18}$$

becomes zero. The parameter values determined by this equation (we denote them by ξ_0^ℓ, λ^ℓ) lie on the boundary of the stability region [1], and therefore, the relation

$$\left|z_i\left(\xi_0^\ell, \lambda^\ell\right)\right| = 1$$

is valid, whereas the inequalities should hold: $|z_i| < 1$.

Therefore, we do not aim at investigating the modes corresponding to these parameter values.

Assuming random changes R delta correlated that corresponds to independent random deviations R at the moment j—the impact from all previous impacts, then the correlation function will have the form:

$$\overline{\Delta R_i \Delta R_j} = \delta_{ij} \overline{\Delta R^2} \tag{5.1.18}$$

where $\overline{\Delta R^2}$ is independent of i and j [3].

Using correlation functions (5.1.18), we can derive analytical expressions for mean square deviations $\overline{\Delta R^2}$ and $\overline{\Delta\tau^2}$. Thus, from relations (5.1.15) for mean squares we will have:

$$\overline{\Delta\tau_n^2} = \frac{\upsilon^2}{(z_1 - z_2)^2}\sum_{\kappa,\nu=-\infty}^{n-1}\left(z_1^{n-\nu-1} - z_2^{n-\nu-1}\right)\left(z_1^{n-\kappa-1} - z_2^{n-\kappa-1}\right)\overline{\Delta R_{\kappa+1}\Delta R_{\nu+1}} \tag{5.1.18}$$

It follows from (5.1.18) that the correlation function $\overline{\Delta R_{\kappa+1}\Delta R_{\nu+1}}$ is other than zero only for $\kappa = \nu$, and therefore, in equality (5.1.18), we pass from the double sum to the single one and taking into account relations (5.1.18), in which the coefficient at δ_{ij} is independent of indices i, j we obtain:

$$\overline{\Delta\tau_n^2} = \frac{\upsilon^2\overline{\Delta R^2}}{(z_1 - z_2)^2}\sum_{\kappa=-\infty}^{n-1}\left(z_1^{n-\kappa-1} - z_2^{n-\kappa-1}\right)^2$$

Performing elementary summation, expression for $\overline{\Delta\tau_n^2}$ will be written in the form:

$$\overline{\Delta\tau_n^2} = \frac{(1 + z_1 z_2)\upsilon^2\overline{\Delta R^2}}{(1 + z_1 z_2)\left(1 - z_1^2\right)\left(1 - z_2^2\right)} = T^2 C_{11}\overline{\Delta R^2} \tag{5.1.19}$$

When calculating $\overline{\Delta y_n^2}$ we do the same

$$\overline{\Delta y_n^2} = \frac{\dot{\xi}_0^2}{(z_1 - z_2)^2}\sum_{\kappa,\nu=-\infty}^{n-1}\left(z_1^{n-\nu-1} - z_2^{n-\nu-1}\right)\left(z_1^{n-\kappa-1} - z_2^{n-\kappa-1}\right)$$
$$\overline{\left(\Phi_1\Delta R_{\nu+1} - \Delta R_{\nu+2}\right)\left(\Phi_1\Delta R_{\kappa+1} - \Delta R_{\kappa+2}\right)} \tag{5.1.20}$$

The difference between these sums and (5.1.18) is that we have to perform a summation of the form:

$$\Lambda = \sum_{\kappa,\nu=-\infty}^{n-1} z_1^{n-\nu-1} z_1^{n-\kappa-1}\overline{\Delta R_{\kappa+2}\Delta R_{\nu+1}}$$

If in (5.1.20) we substitute $\overline{\Delta\tau_n^2}$, then further summation will be carried out in the same way as in (5.1.18), as a result of which (5.1.20) can be rewritten in the form

$$\Lambda = \sum_{\nu=-\infty}^{n-1}(z_1 z_2)^{n-\nu}\overline{\Delta R^2} + \sum_{\nu=-\infty}^{n-1} z_2^{n-\nu-1}\overline{\Delta R_{n+1}\Delta R_{\nu+1}}$$

The second term is equal to zero, and finally, we obtain:

$$\Lambda = \frac{z_1 z_2}{(1 - z_1 z_2)} \overline{\Delta R^2}$$

After some transformations, the standard deviation of impact velocity can be written as follows

$$\overline{\Delta y_n^2} = k\dot{\xi}_0^2\left[(1 + \Phi_1^2)u - 2\Phi_1 w\right]\overline{\Delta R^2} = \dot{\xi}_0^2 C_{12}\overline{\Delta R^2} \tag{5.1.20}$$

where the following notations are introduced

$$\begin{aligned} k &= \left[(1 - z_1 z_2)(1 - z_1^2)(1 - z_2^2)\right]^{-1} \\ u &= 1 + z_1 z_2,\ w = z_1 + z_2 \end{aligned} \tag{5.1.21}$$

For simplicity, we consider the case $\lambda = 0$, corresponding to the spring less single-mass vibro-striker considered in [4]. We rewrite the point mapping (5.1.8) as follows:

$$\begin{aligned} 0 &= y_n(\tau_{n+1} - \tau_n) - \cos\tau_{n+1} + \cos\tau_n - (\tau_{n+1} - \tau_n)\sin\tau_n \\ &\quad - \frac{p}{2}(\tau_{n+1} - \tau_n)^2 \\ y_{n+1} &= -\left[y_n + \sin\tau_{n+1} - \sin\tau_n - p(\tau_{n+1} - \tau_n)\right]R_{n+1} \end{aligned} \tag{5.1.22}$$

where p is the ratio of the weight of the striker to the amplitude of the driving force. Note that, preserving the notation of [4], the point mapping (5.1.22) is obtained from the condition that the external effect has the form $F_0 \cos \omega t$.

Linearizing (5.1.22) in the neighborhood of a stable periodic motion, we obtain a system of difference equations in the form:

$$\begin{aligned} \Delta\tau_{n+1} &= \left\{1 + \sqrt{1 - \left(\frac{1+R}{p}\right)^2 - (1-R)^2(\pi n)^2}\right\}\Delta\tau_n + \frac{1+R}{p}\Delta y_n \\ \Delta y_{n+1} &= \left\{\left[p + \sqrt{1 - \left(\frac{1-R}{1+R}\right)^2 (\pi n p)^2}\right]\right. \\ &\quad \left.\left[\sqrt{\left(\frac{1+R}{p}\right)^2 - (1-R)^2(\pi n)^2}\right] R\right\}\Delta\tau_n \\ &\quad + \left\{\left[R + \frac{1+R}{p}\sqrt{1 - \left(\frac{1-R}{1+R}\right)^2 (\pi n p)^2}\right] R\right\}\Delta y_n \\ &\quad + \frac{2\pi n p}{1+R}\Delta R_{n+1} \end{aligned} \tag{5.1.23}$$

Solving (5.1.23) with respect to P, we obtain

$$\tau_{n+2} = \left[\left(1 - R^2\right) - \frac{(1+R)^2}{p}\cos\tau^*\right]\Delta\tau_{n+1} - R^2\Delta\tau_n + T\Delta R_{n+1} \tag{5.1.24}$$

where $\sin\tau^* = \left[\frac{(R-1)}{(1+R)}\right]\pi np$.

Equation (5.1.24) is a second-order non-uniform difference equation, which can be solved using the method described in solving Eq. (5.1.10). Solving in such a way (5.1.24) and using the stability condition $|z_i| < 1$, we obtain:

$$\Delta\tau_{n+1} = T\sum_{\nu=-\infty}^{n-1}\alpha_{n,\nu}\Delta R_{\nu+1} \tag{5.1.25}$$

where T denotes period $2\pi n$.

Performing the same as for $\Delta\tau_n$, we obtain:

$$\Delta y_n = \frac{Tp}{1+R}\sum_{\nu=-\infty}^{n-1}\alpha_{n,\nu}\left\{\left[1+\sqrt{\left(\frac{1+R}{p}\right)^2 - (1-R)^2(\pi np)^2}\right]\Delta R_{\nu+1} - \Delta R_{\nu+2}\right\} \tag{5.1.26}$$

Assuming the random changes R are delta-correlated, from (5.1.25) and (5.1.26) we obtain the expressions for standard deviations in the form:

$$\begin{aligned}\overline{\Delta\tau_n^2} &= \frac{T^2(1+z_1z_2)}{(1+z_1z_2)\left(1-z_1^2\right)\left(1-z_2^2\right)}\overline{\Delta R^2}\\ \overline{\Delta y_n^2} &= T^2p^2\left[(1+m^2)(1+z_1z_2)\right]\end{aligned} \tag{5.1.27}$$

where m denotes $\left[1 - \frac{1+R}{p}\cos\tau^*\right]$.

From (5.1.25)–(5.1.27) it is clear that neither $\overline{\Delta\tau_n^2}$ nor $\overline{\Delta y_n^2}$ vanish at any parameter value of the system. This follows from the fact that $|z_i| < 1$ and $1 + z_1z_2 = 1 + R^2$. However, you can specify the values of the system parameters at which dispersion $\overline{\Delta y_n^2}$ is zero

$$\begin{aligned}&\left(1+R^2\right)\left[2 + \left(\frac{1+R}{p}\cos\tau^*\right)^2 - 2\frac{1+R}{p}\cos\tau^*\right]\\ &- 2\left[1 - \frac{1+R}{p}\cos\tau^*\right]\left[1 + R^2 - \frac{(1+R)^2}{p}\cos\tau^*\right] = 0\end{aligned} \tag{5.1.28}$$

Or after simple transformations we obtain

$$\cos \tau^* = \frac{\left[2-(1-R)^2\right]\left[1-\left(\frac{1-R}{1+R}\pi np\right)^2\right]}{2p} \tag{5.1.29}$$

At smallest $R \ll 1$, which correspond to almost absolutely inelastic impact, we obtain

$$\frac{1-(\pi np)^2(1-2R)}{\sqrt{1-(\pi np)^2}} = \frac{2R+1-(\pi np)^2(1-2R)}{2p} \tag{5.1.30}$$

equation (5.1.28) at $R = 0$ yields

$$p = \left[1+(\pi n)^2\right]^{-1/2} \tag{5.1.30}$$

From (5.1.30), we obtain that for $n = 1p = 0, 276453$, for $n = 2p = 0, 151653$, and for $n = 3p = 0, 10379$.

Case 2. Small random changes in the gap at constant coefficient of restitution.

When studying vibro-impact systems used as vibratory rammers of soil, gravel, and other materials, it is seen that from impact to impact the changes occur both to the coefficient of restitution and the gap. For example, when compacting the ground, the level of impact is shifted. Therefore, it is of interest to consider the case of small random changes in the gap.

The point mapping for random changes in the gap can be written as:

$$\begin{aligned}
\xi_{n+1} &= \xi_n \cos \lambda(\tau_{n+1}-\tau_n) - \alpha \sin \tau_n \cos \lambda(\tau_{n+1}-\tau_n) + \alpha \sin \tau_{n+1} \\
&\quad - \alpha\lambda^{-1} \cos \tau_n \sin \lambda(\tau_{n+1}-\tau_n) - y_n\lambda^{-1} \sin \lambda(\tau_{n+1}-\tau_n) \\
y_{n+1} &= R_{n+1} \left\{ \begin{array}{l} \xi_n\lambda \sin \lambda(\tau_{n+1}-\tau_n) - \alpha\lambda \sin \tau_n \sin \lambda(\tau_{n+1}-\tau_n) \\ -\alpha \cos \tau_{n+1} + \alpha \cos \tau_n \cos \lambda(\tau_{n+1}-\tau_n) \\ -y_n \cos \lambda(\tau_{n+1}-\tau_n) \end{array} \right\}
\end{aligned} \tag{5.1.32}$$

Assuming, as in case 1, that the random deviations of the gap are small, we can linearize (5.1.32) in the neighborhood of the periodic regime that corresponds to the gap ξ_0, in the neighborhood of which random deviations of the gap occur with an average of zero and a dispersion other than zero.

Linearizing Eq. (5.1.32) in the neighborhood of a stable periodic motion, we obtain

$$\begin{aligned}
\Delta\tau_{n+1} &= \Phi_1\Delta\tau_n + Q_1\Delta y_n + \nu(\Delta\xi_{n+1} - \gamma\Delta\xi_n) \\
\Delta y_{n+1} &= (\Phi_1\Phi_2 + Q_2)\Delta\tau_n + (M + \Phi_2 Q_1)\Delta y_n + R\lambda^2\upsilon\Delta\xi_n \\
&\quad + \Phi_2\nu(\Delta\xi_{n+1} - \gamma\Delta\xi_n)
\end{aligned} \tag{5.1.33}$$

Solving the system (5.1.33) with respect to $\Delta\tau_n$, we obtain the second-order difference equation in the form:

$$\begin{aligned}\Delta\tau_{n+2} = &\left\{\left[\left(1+R^2\right) - \left(1-R^2\right)\sin^2\pi n\lambda - (1+R)^2\lambda^{-2}\cos\lambda\pi n\right]\right.\\ &\left.-v\frac{1+R}{\lambda}\xi_0\sin T\lambda\right\}\Delta\tau_{n+1}\\ &- R^2\Delta\tau_n + v\{-R\Delta\xi_n + (R-1)\cos\lambda T\Delta\xi_{n+1} + \Delta\xi_{n+2}\}\end{aligned} \quad (5.1.34)$$

To find the solution (5.1.34), we again use the condition $|z_i| < 1$, after which we will have:

$$\Delta\tau_n = v\sum_{i=-\infty}^{n-1}\alpha_{n,i}\left[-R\Delta\xi_i + f_0\Delta\xi_{i+1} + \Delta\xi_{i+2}\right] \quad (5.1.35)$$

where f_0 denotes $\gamma(R-1)$.

Similarly, we can obtain from (5.1.33) a second-order difference equation with respect to Δy_n, solution of which will be written in the form

$$\Delta y_n = -Rv\sum_{i=-\infty}^{n-1}\alpha_{n,i}\left[f_1(\Delta\xi_i + \Delta\xi_{i+2}) + f_2\Delta\xi_{i+1}\right] \quad (5.1.36)$$

It follows from (5.1.36) that, with an absolutely inelastic impact, random deviations of the gap do not lead to fluctuations in the impact velocity from a deterministic value. This is a physically obvious fact.

Assuming, as in case 1, random deviations of the gap are delta-correlated, after some transformations we obtain:

$$\begin{aligned}\overline{\Delta\tau_n^2} &= \kappa v^2\left\{u\left(u + 2R^3 + f_0^2\right) + 2w\left[f_0(1-R) - Rw\right]\right\}\overline{\Delta\xi^2}\\ &= v^2 C_{21}\overline{\Delta\xi^2}\\ \overline{\Delta\, y_n^2} &= \kappa v^2 R^2\left\{u\left[2f_1^2\left(1-R^2\right) + f_2^2\right] + 2w\left[2f_1 f_2 + f_1^2 w\right]\right\}\\ &= v^2 C_{22}\overline{\Delta\,\xi^2}\end{aligned} \quad (5.1.37)$$

Consider the case of absolutely elastic impact with random deviations of the gap. The point mapping is written as:

$$\begin{aligned}\xi_{n+1} = &\ \xi_n\cos\lambda(\tau_{n+1}-\tau_n) - \alpha\sin\tau_n\cos\lambda(\tau_{n+1}-\tau_n)\\ &+ \alpha\sin\tau_{n+1} - \alpha\lambda^{-1}\cos\tau_n\sin\lambda(\tau_{n+1}-\tau_n)\end{aligned} \quad (5.1.40)$$

Linearizing (5.1.40) in the neighborhood of a stable periodic single-impact n—multiple mode, we can see that the deviation of the moment n—th impact from $\tau_0 + 2\pi n$ has the following dependency:

$$\Delta\tau_n = \frac{\gamma\Delta\xi_n - \Delta\xi_{n+1}}{\left[\xi_0 - \alpha\cos\tau_0(1-\cos\lambda T) - \alpha\lambda^{-1}\sin\tau_0\sin\lambda T\right]} \quad (5.1.41)$$

Which directly implies that

$$\overline{\Delta\tau_n^2} = \frac{(1+\gamma^2)\overline{\Delta\xi^2}}{\left[\dot{\xi}_0 - \alpha\cos\tau_0(1-\cos\lambda T) - \alpha\lambda^{-1}\sin\tau_0\sin\lambda T\right]^2} \tag{5.1.42}$$

During investigation of the vibro-impact mechanism, it was assumed for the case of small random deviations of the coefficient of restitution and the gap that prehistory of the system condition did not affect its further behavior, i.e., idealization of changes in delta—correlated parameter was accepted. However, in reality, the processes occurring during the operation of vibro-impact systems are such that the values of the parameters n—th impact are fully determined only through all parameter values on previous $(n-1)$—x impacts. This leads to problems on studying the motions in systems with impact interactions, in which small random deviations of the non-delta parameters are correlated. Using the results of [5], we study fluctuations in a single-mass vibro-impact mechanism (see Fig. 5.1) for the case of small non-delta-correlated random changes of the coefficient of restitution R and the gap ξ_0.

For a single-mass vibro-impact mechanism (see Fig. 5.1), we find the statistic characteristics in two cases:

1. the coefficient of restitution from impact to impact varies randomly, and the correlation function $\overline{\Delta R_i \Delta R_j}$ at $i \neq j$ does not vanish and the gap remaining constant, equal to ξ_0.
2. The reduced gap varies randomly from impact to impact and the coefficient of restitution remaining constant for the entire operation of the mechanism. Correlation function $\overline{\Delta\xi_\ell \Delta\xi_m}$ at $\ell \neq m$ is nonzero.

Considering the first case, in order to find the standard deviations, we use formulas (5.1.15), from which it immediately follows that

$$\overline{\Delta\tau_{n1}^2} = \frac{\upsilon^2}{(z_1-z_2)^2} \sum_{\kappa,\nu=-\infty}^{n-1} \left(z_1^{n-\nu-1} - z_2^{n-\nu-1}\right)\left(z_1^{n-\kappa-1} - z_2^{n-\kappa-1}\right)\overline{\Delta R_{\kappa+1}\Delta R_{\nu+1}}$$

$$= \frac{\upsilon^2}{(z_1-z_2)^2}\left\{\begin{array}{l} \sum\limits_{\kappa=0}^{\infty} z_1^{\kappa} \sum\limits_{m=-\infty}^{n-1} z_1^{2(n-m-1)}\overline{\Delta R_{m+1}\Delta R_{m-\kappa+1}} \\ + \sum\limits_{\kappa=0}^{\infty} z_2^{\kappa} \sum\limits_{m=-\infty}^{n-1} z_2^{2(n-m-1)}\overline{\Delta R_{m+1}\Delta R_{m-\kappa+1}} \\ - \sum\limits_{\kappa=0}^{\infty}\left(z_1^{\kappa}+z_2^{\kappa}\right) \sum\limits_{m=-\infty}^{n-1} (z_1z_2)^{n-m-1}\overline{\Delta R_{m+1}\Delta R_{m-\kappa+1}} \end{array}\right\}$$

Taking into account that $\overline{\Delta R_{m+1}\Delta R_{m-\kappa+1}}$ is independent of index m [5], we can obtain

$$\overline{\Delta\tau_{n1}^2} = \frac{\upsilon^2}{(1-z_1z_2)(z_1-z_2)} \sum_{\kappa=0}^{\infty}\left(\frac{z_1^{\kappa+1}}{1-z_1^2} - \frac{z_2^{\kappa+1}}{1-z_2^2}\right)\overline{\Delta R_m \Delta R_{m-\kappa}} \tag{5.1.43}$$

Proceeding in a similar way for the deviation of after impact velocity from (5.1.15), we have

$$\overline{\Delta y_{n1}^2} = \frac{\dot{\xi}_0^2}{(z_1 - z_2)(1 - z_1 z_2)} \sum_{\kappa=0}^{\infty} \left[\left(1 + \Phi_1^2\right)\left(\frac{z_1^{\kappa+1}}{1 - z_1^2} - \frac{z_2^{\kappa+1}}{1 - z_2^2}\right) \right.$$
$$\left. -2\Phi_1 \left(\frac{z_1^{\kappa+1}}{1 - z_1^2} - \frac{z_2^{\kappa+1}}{1 - z_2^2}\right)\right] \overline{\Delta R_m \Delta R_{m-\kappa}} \quad (5.1.44)$$

Note that for $\kappa = 0$ (the case of delta-correlated changes in the coefficient of restitution), formulas (5.1.43) and (5.1.44) go into (5.1.19) and (5.1.20), respectively.

Consider, for example, a specific correlation function in the form:

$$\overline{\Delta R_{m+1} \Delta R_{m-\kappa+1}} = \begin{Bmatrix} \overline{\Delta\, R^2}, & npu \quad \kappa = 0 \\ 0.5\overline{\Delta\, R^2}, & npu \quad \kappa = 1 \\ 0, & npu \quad \kappa > 1 \end{Bmatrix} \quad (5.1.45)$$

By choosing the correlation function in such a way, we thereby consider a relationship between neighboring impacts, more precisely between the previous and subsequent impacts.

Substituting (5.1.45) into (5.1.43) and (5.1.44) after simple transformations, the expressions for $\overline{\Delta \tau_n^2}$ and $\overline{\Delta y_n^2}$ will be rewritten in the form:

$$\overline{\Delta \tau_{n1}^2} = \kappa \upsilon^2 (u + 0.5w)\overline{\Delta R^2} \quad (5.1.46)$$

$$\overline{\Delta y_{n1}^2} = \kappa \dot{\xi}_0^2 \left[\left(1 + \Phi_1^2\right)(u + 0.5w) - 2\Phi_1(u + 0.5w)\right]\overline{\Delta R^2} \quad (5.1.47)$$

Find the differences $\overline{\Delta \tau_{n1}^2} - \overline{\Delta \tau_n^2}$ and $\overline{\Delta y_{n1}^2} - \overline{\Delta y_n^2}$. For this, using expressions (5.1.46), (5.1.47) and (5.1.19), (5.1.20), we have

$$\overline{\Delta \tau_{n1}^2} - \overline{\Delta \tau_n^2} = 0.5\kappa \upsilon^2 w \overline{\Delta R^2} = \varepsilon_1 \overline{\Delta R^2} \quad (5.1.48)$$

$$\overline{\Delta y_{n1}^2} - \overline{\Delta y_n^2} = 0.5\kappa \dot{\xi}_0^2 \left[\left(1 + \Phi_1^2\right)w - 2\Phi_1 u\right]\overline{\Delta R^2} = \varepsilon_2 \overline{\Delta R^2} \quad (5.1.49)$$

From the formula (5.1.48), it can be seen that non-delta-correlated changes R at $\lambda_\kappa = \frac{(2\kappa+1)}{n}$ will not change the value of the standard deviation for the impact time, the value of which for both delta-correlated and non-delta-correlated changes R equal zero, i.e., $\lambda_\kappa = \frac{(2\kappa+1)}{n}$ together $\lambda_\kappa = \frac{(2\kappa+1)}{n}$. However, it is easy to verify from (5.1.48) and (5.1.19) for parameter values determined from the equation

$$\dot{\xi}_0\left[\left(1 + R^2\right) - (1 - R)^2 \sin^2 \pi n \lambda - (1 + R)^2 \lambda^{-2} \cos^2 \pi n \lambda\right]$$
$$= \lambda^{-1}(1 + R)\xi_0 \sin T\lambda \quad (5.1.50)$$

$\varepsilon_1 = 0$, and $\overline{\Delta\tau_n^2}$ are not equal to zero. This means that the effect of both delta-correlated changes in the coefficient of restitution and non-delta-correlated changes R on the standard deviation in time is the same. Therefore, for the parameter values within a small neighborhood of this curve, when determining the standard deviation in time for the non-delta-correlated case of R changes, formulas (5.1.19) for the delta-correlated process can be used.

For a zero gap, these parameter values are determined by a fairly simple formula:

$$(1+R)^2\lambda^2 - \left(1+R^2\right) = 2Rtg^2\pi n\lambda \tag{5.1.51}$$

For non-delta-correlated random gap changes at constant R using (5.1.35) and (5.1.36), we will have the following relations:

$$\begin{aligned}
\overline{\Delta\tau_{n1}^2} = {} & \frac{\upsilon^2\left(1-R^2\right)^{-1}}{(z_1-z_2)}\sum_{\kappa=0}^{\infty}\left\{\left(1+f_0^2+R^2\right)\left(\frac{z_1^{\kappa+1}}{1-z_1^2}-\frac{z_2^{\kappa+1}}{1-z_2^2}\right)\right. \\
& -2\left[Rf_0\left(\frac{z_1^{\kappa+2}}{1-z_1^2}-\frac{z_2^{\kappa+2}}{1-z_2^2}\right)+R\left(\frac{z_1^{\kappa+3}}{1-z_1^2}-\frac{z_2^{\kappa+3}}{1-z_2^2}\right)\right. \\
& \left.\left.-f_0\left(\frac{z_1^{\kappa}}{1-z_1^2}-\frac{z_2^{\kappa}}{1-z_2^2}\right)\right]\right\}\overline{\Delta\xi_m\Delta\xi_{m-\kappa}} \\
& -\frac{2\upsilon^2 R}{(z_1-z_2)}\sum_{\kappa=0}^{\infty}\left(z_1^{\kappa}-z_2^{\kappa}\right)\overline{\Delta\xi_m\Delta\xi_{m-\kappa-1}}
\end{aligned} \tag{5.1.52}$$

$$\begin{aligned}
\overline{\Delta y_{n1}^2} = {} & \frac{R^2\upsilon^2}{(z_1-z_2)\left(1-R^2\right)}\sum_{\kappa=0}^{\infty}\left\{\left(2f_1^2+f_2^2\right)\left(\frac{z_1^{\kappa+1}}{1-z_1^2}-\frac{z_2^{\kappa+1}}{1-z_2^2}\right)\right. \\
& -2f_1f_2\left[\frac{z_1^{\kappa}\left(1+z_1^2\right)}{1-z_1^2}-\frac{z_2^{\kappa}\left(1+z_2^2\right)}{1-z_2^2}\right] \\
& \left.+2f_1^2\left(\frac{z_1^{\kappa+3}}{1-z_1^2}-\frac{z_2^{\kappa+3}}{1-z_2^2}\right)\right\}\overline{\Delta\xi_m\Delta\xi_{m-\kappa}} \\
& +\frac{2R^2\upsilon^2 f_1^2}{(z_1-z_2)}\sum_{\kappa=0}^{\infty}\left(z_1^{\kappa}-z_2^{\kappa}\right)\overline{\Delta\xi_m\Delta\xi_{m-\kappa-1}}
\end{aligned} \tag{5.1.53}$$

Note that formulas (5.1.52), (5.1.53) for $\kappa = 0$ pass into the corresponding formulas (5.1.37).

Choosing the correlation function in the form (5.1.45), the expressions (5.1.52), (5.1.53) will have the form:

$$\begin{aligned}
& \overline{\Delta\tau_{n1}^2} = \kappa v^2\left\{\begin{array}{l}\left(1+f_0^2+R^2\right)(u+0.5w)-2Rf_0\left[w+0.5\left(w^2-R^2u\right)\right] \\ -2R\left[w^2-R^2u+0.5\left(w^2-R^2(1+u)\right)\right]+2f_0(w+0.5u)\end{array}\right\} \\
& \overline{\Delta\xi^2} = v^2R\overline{\Delta\xi^2}
\end{aligned} \tag{5.1.54}$$

$$\overline{\Delta y_{nl}^2} = \kappa v^2 R^2 \left\{ \begin{array}{l} (2f_1^2 + f_2^2)(u + 0.5w) \\ +2f_1 f_2 (2w + 0.5(u(1 - R^2) + w^2)) \\ +2f_1^2 [(w^2 - R^2 u)(1 + 0.5w) + 0.5w(w^2 - R^2)] \end{array} \right\}$$

$$\overline{\Delta \xi^2} + v^2 R^2 f_1^2 \overline{\Delta \xi^2} \tag{5.1.55}$$

Find the difference $\overline{\Delta \tau_{n1}^2} - \overline{\Delta \tau_n^2}$ and $\overline{\Delta y_{n1}^2} - \overline{\Delta y_n^2}$ in case of a random gap change (similar to formulas (5.1.48) and (5.1.49)):

$$\overline{\Delta \tau_{m1}^2} - \overline{\Delta \tau_n^2} = 0.5\kappa v^2 \{ w(u + f_0^2) - 2R(w^2 - R^2 u)(1 + f_0) \\ +2(R^3 + f_0 u) \}$$

$$\overline{\Delta \xi^2} - v^2 R \overline{\Delta \xi^2} = \varepsilon_3 \overline{\Delta \xi^2} \tag{5.1.56}$$

$$\overline{\Delta y_{n1}^2} - \overline{\Delta y_n^2} = 0.5\kappa v^2 R^2 \left\{ w \left[\begin{array}{l} 2f_1^2 (1 + w^2) + f_2^2 + w^2 \\ +2f_1 (f_2 w - f_1 R^2 u) - R^2 \end{array} \right] \right. \\ \left. +2f_1 f_2 (1 - R^4) \right\}$$

$$\overline{\Delta \xi^2} + v^2 R^2 f_1^2 \overline{\Delta \xi^2} = \varepsilon_4 \Delta \overline{\xi}^2 \tag{5.1.57}$$

The analytical study is greatly simplified if we consider the case of an absolutely inelastic impact $R = 0$. For this case $\overline{\Delta y_{n1}^2} - \overline{\Delta y_n^2} = \varepsilon_4 = 0$, i.e., deviation of the impact velocity at an arbitrary n-th impact from the velocity in the deterministic process is zero. This is obvious from physical point of view.

Dependences (5.1.37) and (5.1.35) will have the form:

$$\overline{\Delta \tau_{n0}^2} = \frac{1 + \gamma^2 - 2w\gamma}{(1 - w^2)\dot{\xi}_0^2} \tag{5.1.58}$$

$$\Delta \tau_{n0} = \frac{1}{\xi_0} \sum_{i=-\infty}^{n-1} \alpha_{n,\mathrm{K}} (\Delta \xi_{\mathrm{K}+2} - \gamma \Delta \xi_{\mathrm{K}+1}) \tag{5.1.59}$$

From (5.1.59), it follows that having choosen the parameter values equal to $\lambda_\kappa = (2\kappa + 1)/2n$ and assuming occasional deviations of the gap are subject to the law $\Delta \xi_{\mathrm{K}+2} = \Delta \xi_{\mathrm{K}+1}$, we obtain that the time deviations at an arbitrary n-th impact will be zero. That is, for these values λ_k the results of the deterministic theory are fully transferred to the case of nondeterministic changes in the gap.

The mean square deviation will be zero for the parameter values determined by the relation:

$$2w\gamma = 1 + \gamma^2 \tag{5.1.60}$$

which is written in a more expanded form

$$\left(1+\cos^2\lambda T\right)=2\cos\lambda T\left\{\cos^2\pi n\lambda\frac{\lambda^2-1}{\lambda^2}-\frac{\xi_0}{\lambda\dot{\xi}_0}\sin\lambda T\right\} \tag{5.1.61}$$

From (5.1.61), it follows that in the case of an absolutely inelastic impact and zero reduced gap only in the intervals over λ, defined by inequalities

$$0<\lambda_\kappa<\sqrt{3-2\sqrt{2}},\cdot\lambda>\sqrt{3-2\sqrt{2}} \tag{5.1.62}$$

such values m, c and ω may exist at which small random deviations of the gap do not affect the mean square characteristic $\overline{\Delta\tau_{n0}^2}$. Note that if the pre-resonance region is considered ($\lambda > 1$) then the possible parameters lie in the intervals $\frac{4\kappa-1}{4n}<\lambda_\kappa<\frac{4\kappa+1}{4n}$,

$$\kappa\geq 1+[1/4+n\sqrt{3+2\sqrt{2}}] \tag{5.1.63}$$

and for the post-resonance region ($\lambda > 1$)

$$\frac{4\kappa+1}{4n}<\lambda_\kappa<\ell_\kappa,\kappa<1+[n\sqrt{3+2\sqrt{2}}-3/4] \tag{5.1.64}$$

where the multiplicity of the period of motion n to the period of external force l_k is determined

$$\ell_{\text{x}}=\min_\kappa\left\{\sqrt{3+2\sqrt{2}},\frac{3+4\kappa}{4n}\right\}$$

To determine the exact values n, it is necessary to solve a transcendental equation of the form:

$$\cos\lambda T=\frac{\left(\lambda^2-1\right)\pm\sqrt{\lambda^4-6\lambda^2+1}}{2} \tag{5.1.65}$$

With an absolutely elastic impact in the case of a non-delta-correlated gap changes, the root mean square characteristics $\overline{\Delta\tau_n^2}$ will have the form:

$$\overline{\Delta\tau_{n1}^2}=\frac{\left[(1-\gamma^2)(1+0.5w)-2\gamma(w+0.5)\right]}{\left(1-w^2\right)\dot{\xi}_0^2}\overline{\Delta\xi^2} \tag{5.1.66}$$

Taking into account (5.1.58) and (5.1.66), the coefficient ε_3 will be written in the form

$$\varepsilon_3=\frac{w\left(1-\gamma^2\right)-2\gamma}{2\dot{\xi}_0^2\left(1-w^2\right)}\overline{\Delta\xi^2} \tag{5.1.67}$$

We choose parameters for the system satisfying the relation

$$2\gamma = w\left(1 - \gamma^2\right) \quad (5.1.68)$$

And find that the correlation at such parameters does not affect the rms characteristics. The relation (5.1.68) will be rewritten in the form

$$2\cos\lambda T = \left(1 + \cos^2\lambda T\right)\left\{\frac{\lambda^2 - 1}{\lambda^2}\cos^2\lambda\left(\frac{T}{2}\right) - \frac{\xi_0}{\lambda\dot{\xi}_0}\sin\lambda T\right\} \quad (5.1.69)$$

If random deviations of the gap occur in the neighborhood of zero value, then to find the values λ_k, in which $\varepsilon_3 = 0$, we obtain transcendental equation of the form

$$2\lambda^2\alpha\cos\lambda T = \left(1 + \cos^2\lambda T\right)\cos^2\lambda(T/2) \quad (5.1.70)$$

from which it directly follows that if the selected parameters correspond to the pre-resonance region, then λ_k is determined from the interval

$$\frac{4\kappa - 1}{4n} < \lambda_\kappa < \frac{4\kappa + 1}{4n}, \kappa \geq 1 + [n + 1/4]$$

and in a post-resonance region, these values lie in

$$\frac{4\kappa + 1}{4n} < \lambda_\kappa < \ell_\kappa, \cdot\kappa = 0, 1, \ldots, 1 + [n - 3/4]$$

where $\ell_\kappa = \min_\kappa\{1, (3 + 4\kappa)/4n\}$

To consider the general case, when in operation of a vibro-impact system both the gap and the coefficient of restitution change from impact to impact, we linearize the point mapping (5.1.7) in the neighborhood of stable periodic motion, whereafter we obtain the system of difference equations in the form

$$\begin{aligned}\Delta\tau_{n+1} &= \Phi_1\Delta\tau_n + Q_1\Delta y_n + v(\Delta\xi_{n+1} - \gamma\Delta\xi_n)\\ \Delta y_{n+1} &= (\Phi_1\Phi_2 + Q_2)\Delta\tau_n + (M + \Phi_2 Q_1)\Delta y_n + R\lambda v\sin\lambda T_\Delta\xi_n - \dot{\xi}_0\Delta R_{n+1}\end{aligned} \quad (5.1.70)$$

which differs from (5.1.33) by the last term in the second Eq. (5.1.70).

Solving the system of difference Eq. (5.1.70) and taking into account that the roots of the characteristic equation modulo less than 1, we obtain

$$\Delta\tau_n = v\sum_{i=-\infty}^{n-1}\alpha_{n,i}\left[-R\Delta\xi_i + f_0\Delta\xi_{i+1} + \Delta\xi_{i+1} + v\dot{\xi}_0\Delta R_{i+1}\right] \quad (5.1.71)$$

$$\Delta y_n = -Rv \sum_{i=-\infty}^{n-1} \alpha_{n,i}\big[f_1(\Delta\xi_i + \Delta\xi_{i+2}) + f_1\Delta\xi_{i+1}$$
$$+\dot{\xi}_0^2 R^{-1}(\Delta\xi_{i+2} - \Phi_1\Delta\xi_{i+1})\big] \quad (5.1.72)$$

If we further assume that random changes of both the gap and the coefficient of restitution are delta-correlated and there is no correlation between the change in the gap and R, then the following relations will be satisfied:

$$\overline{\Delta\xi_i\Delta\xi_K} = \delta_{iK}\overline{\Delta\xi^2},\ \overline{\Delta R_i\Delta R_K} = \delta_{iK}\overline{\Delta R^2},\ \overline{\Delta\xi_i\Delta R_K} = 0s \quad (5.1.73)$$

Using the correlation functions (5.1.73), it is easy to obtain expressions for the rmv values in the form

$$\overline{\Delta\tau_n^2} = T^2 C_{11}\overline{\Delta R^2} + v^2 C_{21}\overline{\Delta\xi^2} \quad (5.1.74)$$

$$\overline{\Delta y_n^2} = \dot{\xi}_0 C_{12}\overline{\Delta R^2} + v^2 C_{22}\overline{\Delta\xi^2} \quad (5.1.75)$$

From (5.1.74) and (5.1.75), it is clearly seen that the root mean square characteristics of the system for the case of a simultaneous random change in the gap and the coefficient of restitution upon impact are equal to the sum of the previously found rms values.

The first term in (5.1.74) and (5.1.75) is the mean square for the coefficient of restitution varying at a constant gap, and the second term is the rms value for gap changing at a constant R. Thus, we may conclude that in order to take into account the effect of random deviations of both the gap and the coefficient of restitution, it is necessary to use the results obtained when considering random deviations of the gap (coefficient of restitution $R/$ at constant R (reduced gap). This is a natural conclusion for Markov processes.

If random deviations R are non-delta correlated, i.e., equalities (5.1.73) are not satisfied, then such a conclusion cannot be drawn. The rms deviations will be cumbersome and difficult to analyze. Note that $\overline{\Delta\tau_n^2}$ will consist of three terms, two of which are derived from formulas (5.1.43), (5.1.44) and (5.1.52), (5.1.53), and the third term is a cumbersome expression, which includes the correlation function of the gap R of the coefficient of restitution (ξ_0).

5.2 Investigation of Stochastic Stability

To study the dynamics of real-time operation vibro-impact systems, it is necessary to study the problems of the behavior of randomly perturbed dynamic systems. The main issue in studying the dynamics of such systems is investigation of stability.

To study the stability of the systems in the presence of random perturbations, Katz et al. in [6–15] and other researchers introduced statistical generalizations of the concept of stability such as Lyapunov stability and considered the main theorems of the direct method.

So, under the stability of a trivial solution of a differential equation $\frac{dx}{dt} = G(x, t, \xi(t, \alpha))$, we mean the following definitions [6]:

1. $x(t) \equiv 0$ is stable (weak) with probability at $x(t) \equiv 0$, if for any $\varepsilon > 0$, we find such $\delta > 0$, that at $t \geq t_0, \cdot |x_0| < rP\{|x(t, \alpha, t_0, x_0)| > \varepsilon\} < \delta$
2. $x(t) \equiv 0$ p is stable if for any $\varepsilon > 0$ we find such $r > 0$ that at $t \geq t_0, \cdot |x_0| < r$ $M\,|x(t, \alpha, t_0, x_0)|^p < \varepsilon, (p > 0)$
3. $x(t) \equiv 0$ is exponentially stable if there are constants $A > 0$ and $\gamma > 0$ such that $M\,|x(t, \alpha, t_0, x_0)|^p \leq A, |x_0|^p \exp\{-\gamma(t - t_0)\}$ and others.

However, the results obtained using these concepts of stability mainly relate to linear systems or systems close to linear.

Vaidenhammer in [16] finds the stability conditions for the equilibrium position of a relatively simple linear system with two degrees of freedom, parametrically perturbed by white noise. The definition of stability was introduced using the results of the correlation theory as follows: a system is considered stable if the first- and second-order statistical moments of coordinates and velocities remain limited over time.

Based on general considerations given in the work of Neymark [17], V. A. Brusin and M. L. In [13], we introduce the concept of stochastic stability, which can be adopted for essentially nonlinear systems. So, a dynamic system, the input of which is a random process $\xi(\lambda, t)$ and exit-random process $y(\lambda, t)$, in [13] is called stochastically stable if for any $\varepsilon > 0$ there is such $\delta > 0$, that if $r(\varepsilon) < \delta$, then $\rho(y) \leq \varepsilon(\delta)$ and $\varepsilon \to 0$ at $\delta \to 0$. The concept of stochastic stability depends on the way standards $r(\varepsilon)$ and $\rho(y)$ are introduced.

Using all of the aforesaid, we extend these concepts of stochastic stability to the considered single-mass vibro-impact systems. However, we note that everything we say about the systems with one degree of freedom will automatically extend to systems with many degrees of freedom.

In Sect. 5.1 it was shown that the standard deviations $\overline{\Delta\tau_n^2}$ and $\overline{\Delta y_n^2}$ can be written as the product of a certain constant, which is a function of only the system parameters, by the variance of random deviations either $\overline{\Delta R^2}$ or $\overline{\Delta\xi^2}$ depending on which parameter R or ξ_0 changes randomly.

Suppose that in a system with a constant gap ξ_0 the coefficient of restitution R changes randomly from impact to impact and $\overline{\Delta R^2} = 0$ while $\overline{\Delta R^2} \neq 0$. Then, using [13], we can assume that the system will be stochastically stable if there exists a constant C, depending only on the parameters of the system, such that for sufficiently large n the inequalities held (5.2.1).

$$\overline{\Delta\tau_n^2} + \overline{\Delta y_n^2} \leq C\overline{\Delta R^2} \quad (5.2.1)$$

We explain inequalities (5.2.1). Since all the results obtained in § 5.1 are associated with small changes in R and ξ_0, and the deviation of the parameters occurs in the vicinity of the values, therefore, in the deterministic case, a stable periodic regime exists. Since the value $\overline{\Delta R^2}$ is small, then the expression on the right side of the inequality is also small (C is constant). It is clear that in inequality (5.2.1) due to the smallness of $C\overline{\Delta R^2}$, rms values $\overline{\Delta\tau_n^2}$ and $\overline{\Delta y_n^2}$ should be also small. It also follows from (5.2.1) that if the deviations of the coefficient of restitution tend to zero, then the mean squares also tend to zero if the system is stochastically stable.

Similarly, to inequality (5.2.1), we determine the stochastic stability of the system for the case when the gap randomly changes and the coefficient remains constant. So, we assume that the system is stochastically stable under random deviations of the gap with an average zero value and a dispersion other than zero if the inequality holds

$$\overline{\Delta\tau_n^2} + \overline{\Delta y_n^2} \leq C\overline{\Delta\xi^2} \tag{5.2.2}$$

The meaning of inequality (5.2.2) is similar to that of (5.2.1).

By constant values C, and therefore, by the values of the constants C_{ij} $i,\ j= 1, 2$, you can judge the system stability degree. In Fig. 5.2a, b in the stability region found in [1] for a deterministic process for values λ, less than 1, level lines for $n= 1$, $R= 0, 6$, values C_{11}, C_{12} are constructed, respectively. Note that for the values $\lambda > 1$ a significant part of the stability region is cut off by the bifurcation boundary C_τ, and therefore, level lines in these areas are not given. Dashed line shows bifurcation curve C_τ. It is seen from Fig. 5.2 shows that the smallest coefficient C is in the vicinity $\lambda = 0.5$, thus choosing λ in the vicinity of this value, and the effect of random changes in the coefficient of restitution will be negligible. If you plot level lines in stability areas for $\lambda > 1$, then the least value will be in a neighborhood of values λ, determined from equality $\lambda_\kappa = \frac{2\kappa+1}{2n}$.

Approaching the boundaries of the stability region, the value of C increases to infinity on the boundary itself; however, the case of the stability boundary is beyond the scope.

Qualitative course of level lines C_{21},C_{22} is similar to that of C_{11}, C_{12}.

To determine the quantitative characteristics of the effect of correlation functions on rms deviations when changing R, we consider for simplicity the motion of spring less vibro hammer [4].

In Fig. 5.3, dashed lines correspond to level lines C_1 with delta-correlated changes R, and solid lines show the level C_1^0 with non-delta-correlated changes R.

There is a delimiting curve inside the stability region

$$x = \frac{2(1+R)^2}{\sqrt{4\left(1+R^2\right)^2 + \left(1-R^2\right)^2 T^2}} \tag{5.2.3}$$

the meaning of which is as follows.

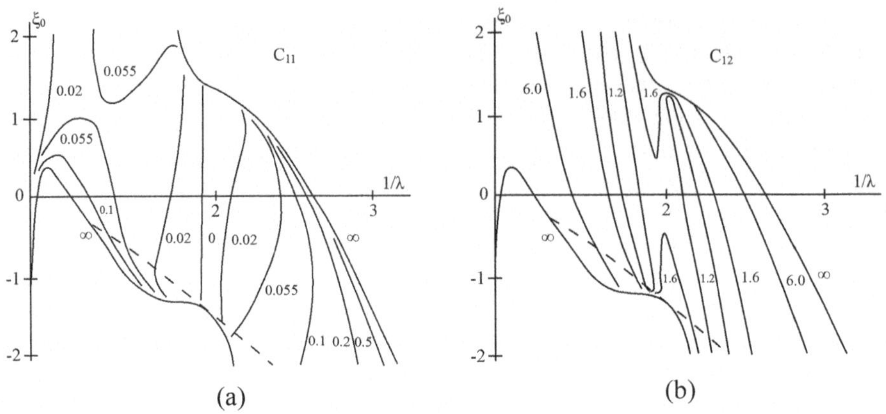

Fig. 5.2 **a** In the stability area, level lines were constructed for $n = 1$, $R = 0.6$. **b** In the stability area, level lines were constructed for $n = 1$, $R = 0.6$

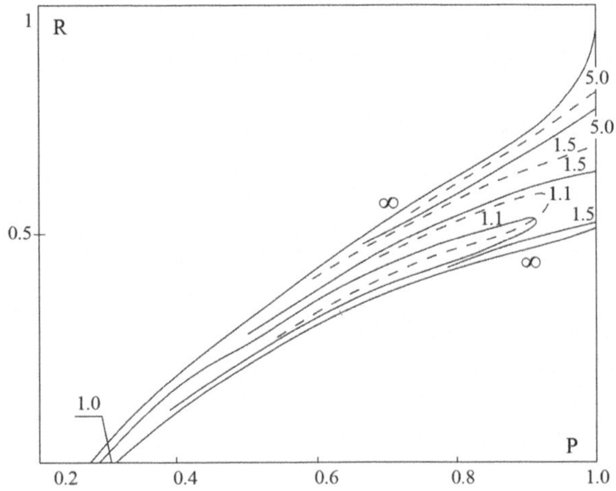

Fig. 5.3 Dashed lines show level lines C_1 for delta-correlated changes R, and solid lines C_0^1 show level lines R corresponding to week-correlated changes

Random changes R when accounting lead to level lines displacement to smaller p for $p \leq x$ and to the area of large p for $p > x$. It should be noted that with $p > x$ the probability of a system exiting a given mode is extremely high, since the values C_1 and C_1^0 almost go to infinity. Based on this, when choosing the parameters of the system, it is necessary to be guided by the inequality $p < x$.

Similarly as in Fig. 5.3 will behave level lines for values C_2 and C_2^0.

5.3 The Influence of Random Deviations of the Coefficient of Restitution *R* by the Average Optimal Impact Velocity

When tuning vibro-impact systems, the conditions of either the maximum impact pulse or maximum impact velocity are usually used as the criterion of optimality. According to this, the system parameters are selected in the vicinity of those parameter values at which the maximum impact pulse (impact velocity) is reached.

Due to fluctuations in the coefficient of restitution and the gap, the position and magnitude of the maximum impact velocity will shift. Therefore, an important theoretical and practical task arises of finding this new position and a new value for the maximum impact velocity.

Assume random deviations R occur in the neighborhood $R = R_0$ with a mean zero value and a dispersion other than zero. We also assume that the deviations R are small.

In order to find the average value of the impact velocity, it is impossible to fully use only the results obtained in Chaps. 1, 2, since if we find the average velocities, they will be zero due to the condition $\overline{\Delta R} = 0$.

Since we will continue to be interested in the average value of impact velocity, the point mapping in this case can be written in the following form:

$$\begin{aligned}
\sin \tau_{n+1} &= \xi_0 \alpha^{-1}\left[1 - \cos \lambda(\tau_{n+1} - \tau_n)\right] + \sin \tau_n \cos \lambda(\tau_{n+1} - \tau_n) \\
&\quad + \lambda^{-1} \cos \tau_n \sin \lambda(\tau_{n+1} - \tau_n) + R_n \lambda^{-1} \alpha^{-1} \xi_n \sin \lambda(\tau_{n+1} - \tau_n) \\
\xi_{n+1} &= \alpha \cos \lambda \tau_{n+1} - \xi_0 \lambda \sin \lambda(\tau_{n+1} - \tau_n) + \lambda \alpha \sin \tau_n \sin \lambda(\tau_{n+1} - \tau_n) \\
&\quad - \alpha \cos \tau_n \cos \lambda(\tau_{n+1} - \tau_n) - R_n \dot{\xi}_n \cos \lambda(\tau_{n+1} - \tau_n)
\end{aligned} \tag{5.3.1}$$

Where $\dot{\xi}_n$ denote the velocity directly before n-th impact.

We expand the point mapping (5.3.1) in a series by $\Delta\tau_n$ and $\Delta\dot{\xi}_n$ up to the second order inclusively, and, citing similar terms, we obtain a nonlinear system of difference equations

$$\begin{aligned}
\Delta\tau_{n+1} &= a_1 \Delta\tau_n - R Q_1 \Delta\dot{\xi}_n + v \Delta R_n + \alpha(n) \\
\Delta\dot{\xi}_{n+1} &= b_1 \Delta\tau_n + f_1 \Delta\tau_{n+1} - \gamma \dot{\xi}_0 \Delta R_n + M \Delta\dot{\xi}_n + \rho(n)
\end{aligned} \tag{5.3.2}$$

in which through

$$\alpha(n) = \sum_{i=1}^{8} c_i x_i, \cdot \rho(n) = \sum_{i=1}^{8} g_i x_i \tag{5.3.3}$$

second-order quantities are denoted with respect to $\Delta\tau_n$ and $\Delta\dot{\xi}_n$, and functions c_i and g_i and x_i are defined as follows

$$\begin{aligned}
&x_1 = \Delta\tau_n\Delta\tau_{n+1}, x_2 = \Delta\tau_n\Delta R_n, x_3 = \Delta\tau_n\Delta\dot{\xi}_n, x_4 = \Delta\tau_n\Delta R_n \\
&x_5 = \Delta\tau_{n+1}\Delta\dot{\xi}_n, x_6 = \Delta R_n\Delta\dot{\xi}_n, x_7 = \Delta\tau_n^2, x_8 = \Delta\tau_{n+1}^2 \\
&c_1 = 2c_2c_8, c_2 = -c_4 = g_6 = -\gamma, c_3 = -c_5 = M\xi_0^{-1}, c_6 = -Q_1 \\
&c_7 = 0.5\left(\nu c_4\dot{\xi}_0^{-1} - c_1\right), c_8 = -0.5f_1\xi_0^{-1}, g_1 = -g_5 = -R\lambda^2\nu \\
&g_7 = 0.5(c_2\cos\tau_0 - g_1), g_8 = 0.5\left(\cos\tau_0 - g_1\lambda^2\xi_0\right) \\
&b_1 = -\left(f_1\gamma + R\dot{\xi}_0\lambda^2\nu\right), a_1 = M - fQ, b_2 = f_1
\end{aligned} \tag{5.3.4}$$

Excluding $\Delta\tau_n$ from (5.3.2), for $\Delta\dot{\xi}_n$ we drive the difference equation in the form

$$\begin{aligned}
\Delta\dot{\xi}_{n+2} &= \alpha_1\Delta\dot{\xi}_{n+1} + \beta_1\Delta\dot{\xi}_n - R\dot{\xi}_0\Delta R_n + \Phi_1\Delta R_{n+1} + \\
&+ b_1\alpha(n) - a_1(\rho(n) + b_2\alpha(n+1) + \rho(n+1))
\end{aligned} \tag{5.3.5}$$

where α_1 and β_1 see in (5.1.14).

By taking an ensemble average [18] the difference Eq. (5.3.5) we obtain:

$$\begin{aligned}
\overline{\Delta\dot{\xi}_{n+2}} &= \alpha_1\overline{\Delta\dot{\xi}_{n+1}} + \beta_1\overline{\Delta\dot{\xi}_n} + b_1\overline{\alpha(n)} - a_1\overline{\rho(n)} \\
&+ b_2\overline{\alpha(n+1)} + \overline{\rho(n+1)}
\end{aligned} \tag{5.3.6}$$

or accounting for the properties of the functions

$$\overline{\rho(n)} = \overline{\rho(n+1)}, \overline{\alpha(n)} = \overline{\alpha(n+1)} \tag{5.3.7}$$

Equation (5.3.6) is rewritten in the form

$$\overline{\Delta\dot{\xi}_{n+2}} = \alpha_1\overline{\Delta\dot{\xi}_{n+1}} + \beta_1\overline{\Delta\dot{\xi}_n} + (b_1 + f_1)\overline{\alpha(n)} + (1 - a_1)\overline{\rho(n)} \tag{5.3.8}$$

The right-hand side of (5.3.8) contains nonlinear terms with respect to $\Delta\tau_n$ and $\Delta\dot{\xi}_n$. In order to solve (find the second approximation), we do the following: using the results obtained in 1, namely formulas (5.2.5), we calculate the necessary moments in the functions $\overline{\rho(n)}$ and $\overline{\alpha(n)}$. So, after some transformations we have:

$$\begin{aligned}
\overline{\Delta\tau_n\Delta R_n} &= 0 = t_2 \\
\overline{\Delta\tau_n\Delta\dot{\xi}_n} &= \nu\kappa\dot{\xi}_0(\Phi_1 u - Rw)\overline{\Delta R^2} = t_1\overline{\Delta R^2} \\
\overline{\Delta\tau_n\Delta\tau_{n+1}} &= \nu^2\kappa w\overline{\Delta R^2} = t_7\overline{\Delta R^2} = t_8\overline{\Delta R^2} \\
\overline{\Delta\tau_{n+1}\Delta R_n} &= \nu\overline{\Delta R^2} = t_4\overline{\Delta R^2} \\
\overline{\Delta\tau_n\Delta\dot{\xi}_n} &= \nu\kappa\dot{\xi}_0\left[w\Phi_1 - R\left(w^2 - R^2u\right)\right]\overline{\Delta R^2} \\
&= t_5\overline{\Delta R^2}, \overline{\Delta\dot{\xi}_n\Delta R_n} = 0 = t_6
\end{aligned} \tag{5.3.9}$$

where $\Phi_1 = \left[\left(\sin\tau_0 - \lambda^2\dot{\xi}_0^{-1} - \cos\lambda T\right)\right]$

Substituting in $\overline{\alpha(n)}$ and $\overline{\rho(n)}$ expressions (5.3.9) and solving the resulting difference equation, provided $|z_i| < 1$, we obtain:

$$\overline{\Delta\dot{\xi}_{n+2}} = \left[(b_1 + f_1)\Phi_1 - (1 - a_1)\Phi_2\right]\overline{\Delta R^2} = \overline{E\Delta R^2} \tag{5.3.10}$$

where the following notations are introduced

$$\Phi_1 = \sum_{i=1}^{8} c_i t_i,\ \Phi_2 = \sum_{i=1}^{8} g_i x_i \tag{5.3.11}$$

To determine the effect of random deviations R on the location of the optimal impact velocity, we construct in the plane of the parameters λ^{-1}, ξ_0 the lines of coefficient level E for $R = 0.6$, located in the area of sustainability.

In Fig. 5.4, solid lines show level lines E, and a dotted line corresponds to the line with maximum value $\dot{\xi}_0$, for a deterministic case. Note that the maximum is found with a fixed gap in the variable λ, i.e.,

$$\max_{\lambda\in[a,b]} \dot{\xi}_0(\lambda, \dot{\xi}_0)$$

Figure 5.4 shows that in the plane λ^{-1}, ξ_0 there exist domain D, bounded by curve 1 and line $\lambda = 0.5$, inside which E is negative, but outside-positive. On curve 1 and line $\lambda = 0.5$, i.e., on the border of this region, the coefficient value equals zero. The line with maximum impact velocity, $\dot{\xi}_0$, lies inside the domain D. It follows that with random deviations R the average value of the optimal velocity decreases. This figure also shows that with a constant gap, the average value $\overline{\Delta\dot{\xi}_n}$ inside the area D less than or equal to the velocity in the deterministic case, but outside the region D more than that value. For parameter values on curve 1 and line $\lambda = 0.5$, random changes R do not affect the average impact velocity,i.e., the following ratio is satisfied:

$$\overline{\Delta\dot{\xi}_n} = 0$$

In Fig. 5.5(a, b), solid lines show the dependences of the impact velocity on λ with a constant gap for the deterministic case, and the dashed lines correspond to the curves of the average impact velocity taking into account random deviations R for $n = 1$, $R = 0.6$, and $\overline{\Delta R^2} = 0.5$.

As noted above, the maximum value of the impact velocity decreases, and the maximum shifts with positive gaps toward large λ, and for negative gaps toward smaller λ. In view of this, to maintain the regime in real conditions, the frequency of the driving force ω should be decreased with gaps greater than 0 and increased for gaps less than 0.

Studies with small random deviations of the coefficient of restitution upon impact make it possible to qualitatively evaluate the effect of deviations R by the magnitude

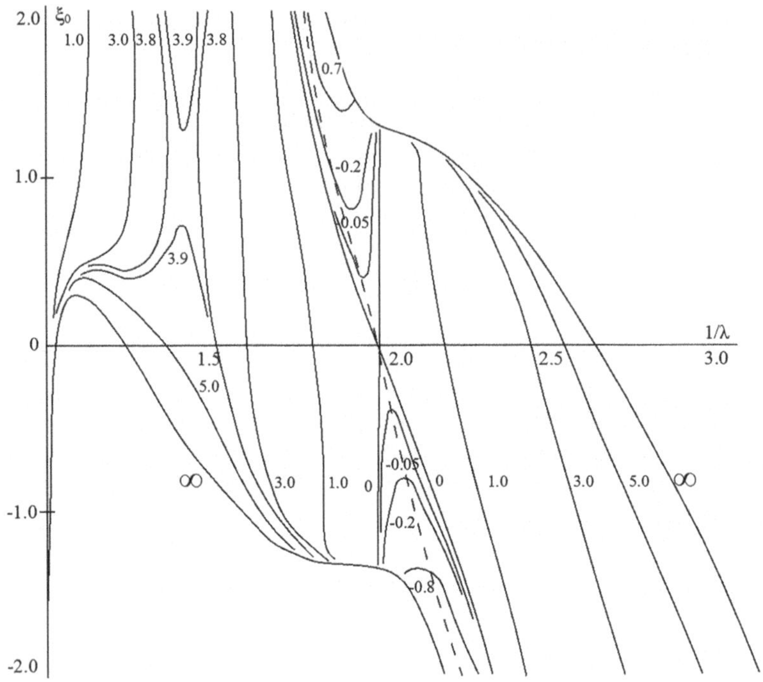

Fig. 5.4 Solid lines show the level lines E, and the dotted line shows the line on which the maximum value lies $\dot{\xi}_0$, corresponding to the deterministic case

of the optimal velocity (by the values of the optimal parameters) and use these qualitative results in real systems.

5.4 Investigation of Probabilities of a System Exit from a Single-Impact Mode of Motion in Case of Random Changes in the Coefficient of Restitution *R*

In the previous section, a theoretical study of the influence of small random deviations R by the average optimal impact velocity. Another practically important issue for tuning vibro-impact systems to the optimal mode is finding such parameter values at which the probability of exit (we denote it by P_b) of the system from a given periodic mode would be minimal. As you know, single-impact mode is considered the most optimal, and therefore, further research will be carried out for $n = 1$ and $m = 1$ (see [1]).

Consider, without going into specific calculations, the general idea of further research to find

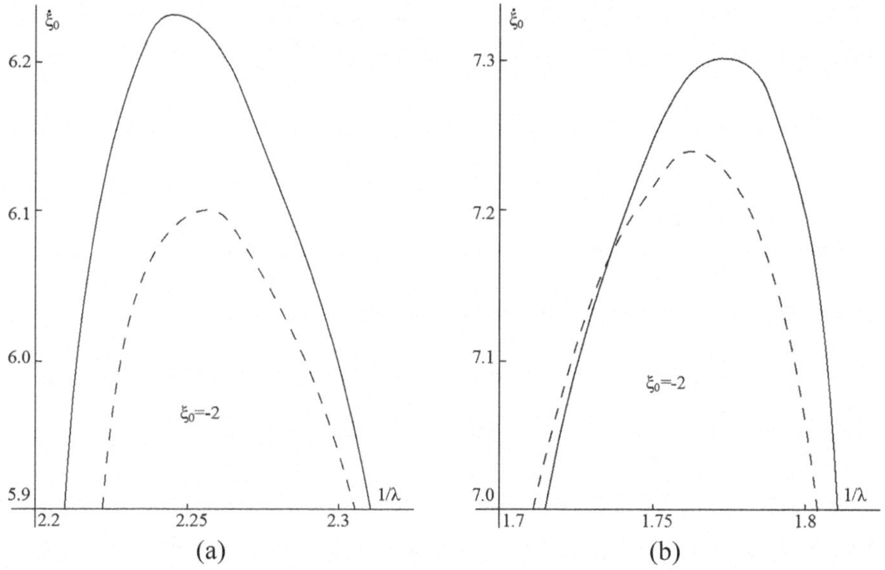

Fig. 5.5 **a** Solid lines show the dependence of the impact speed on λ at a constant gap for the deterministic case, and the dotted lines show the curves of the average impact speed taking into account random deviations R for $n = 1, R = 0.6$ and $\overline{\Delta R^2} = 0.1$. **b** Solid lines show the dependence of the impact speed on λ at a constant gap for the deterministic case, and the dotted lines show the curves of the average impact speed taking into account random deviations R for $n = 1, R = 0.6$ and $\overline{\Delta R^2} = 0.1$

$$P_b^* = \min_j \left\{P_b^j(\alpha_k)\right\}, k = 1, 2, \ldots, N \tag{5.4.1}$$

where α_k is dimensionless parameters of the vibration-impact system.

Let the point transformation of the system under consideration with impact interactions be written as

$$\begin{aligned} x_j^1 &= f_i\left(x_1^0, x_2^0, \ldots, x_k^0, \tau^0, \tau^1, \alpha_1, \ldots, \alpha_n\right) \\ &\varphi\left(x_1^0, \ldots, x_k^0, \tau^0, \tau^1, \alpha_1, \ldots, \alpha_n\right) \end{aligned} \tag{5.4.2}$$

where $\varphi\left(x_1^0, \ldots, x_k^0, \tau^0, \tau^1, \alpha_1, \ldots, \alpha_n\right)$ is the equation of the geometric image on which the blow occurs.

Then the equations for finding the fixed points of the point transformation corresponding to single-impact periodic motions are obtained from (5.4.2) by adding the periodicity conditions

$$x_j^1 = x_i^0, \tau^1 = \tau^0 + \mathrm{T} \tag{5.4.3}$$

Suppose the system of equations for finding fixed points can be easily resolved, and the coordinates of invariant points can be obtained M^* in parametric form

$$x_i^* = g_i(\alpha_1, \ldots, \alpha_n)$$
$$\tau_0 = g_0(\alpha_1, \ldots, \alpha_n) \tag{5.4.4}$$

By assuming further that in the parameter space α_i we find those values for which the periodic motion is stable. In other words, we managed to isolate stability regions in the parameter space; we denote them by D_{11}^i/ the index i means that the stability region, generally speaking, is not simply connected /.

Take some fixed set of system parameters $\alpha_{k_1}, \ldots, \alpha_{k_N}$, and such that $\{\alpha_{k_i}\} \in D_{11}^i$. To this set in phase space some fixed point M_k^* will correspond. We construct (in some way) in the space of initial values x_i^0, τ^0 fixed point attraction domains M_k^*, and we denote them by E_n^j (index j means that the domain of attraction may not be simply connected) see, for example, [19]. Let the point M_K^* belongs to the area E_n^k. Let us plot in this area E_n^k hypersphere centered at a point M_k^* and radius r, equal to the minimum distance from the border of the region E_n^k to the point M_k^*, the equation of which will be

$$\sum_{i=1}^{k} \left(x_i^0 - x_i^*\right)^2 + \left(\tau_i^0 - \tau_i^*\right)^2 = r^2 \tag{5.4.5}$$

Let one or more system parameters α_0^j randomly change from impact to impact

$$\alpha_n^j = \alpha_0^j + \Delta\alpha_n^j \tag{5.4.6}$$

where α_0^j denote parameter values for which in the deterministic case there is a stable periodic one-impact mode. We also assume that random deviations $\Delta\alpha_n^j$ are so small that they do not remove the system from the vicinity of a stable periodic mode and the average $\overline{\Delta\alpha_n^j} = 0$, and dispersions $\overline{(\Delta\alpha_n^j)^2}$ are nonzero.

Linearizing (5.4.2) in a neighborhood of the stability of periodic motion, we obtain a system of difference equations with respect to the coordinates and times of impacts. Solving this system of difference equations, we obtain:

$$\Delta x_i^n = \Phi_i\left(\alpha_0^1, \ldots, \alpha_0^k, \tau^0, \Delta\alpha_n^1, \ldots, \Delta\alpha_n^k\right)$$
$$\Delta\tau^n = \Phi\left(\alpha_0^1, \ldots, \alpha_0^k, \tau^0, \Delta\alpha_n^1, \ldots, \Delta\alpha_n^k\right) \tag{5.4.7}$$

where the functions Φ_i and Φ are solutions to the system of N difference equations.

Consider a random function of the form

$$\xi^2 = \sum_{i=1}^{N} \left(\Delta x_i^n\right)^2 + \left(\Delta\tau^n\right)^2 \tag{5.4.8}$$

which characterizes random deviations in coordinates and in time from a fixed point M_k^*.

We assume that if a random function ξ, left the hypersphere of radius r, then the failure of a single-impact motion occurred. We define an upper bound for this exit probability P_b using the well-known Chebyshev inequalities. This inequality gives an upper bound for the probability that the absolute deviation of the $|x - \xi|$ random variable x from its distribution center $\xi = Mx$ will exceed a given number a, in the form

$$P_b\{|x - \xi| \geq a\} \leq \frac{\sigma^2}{a^2} \tag{5.4.9}$$

where σ^2 is dispersion of a random variable x.

If the laws of distribution of random variables α_j are given, then by the formula (5.4.9), we can find the probability of exit from a given single-impact mode. With other parameter values $\alpha_j^1, \ldots, \alpha_j^l$, we get $P_b^1, P_b^2, \ldots, P_b^l$. Choosing from this totality $P_b^* = \min_l \{P_b^l\}$, we get the parameter values at which the probability of exit is the smallest.

Note that the calculation P_b^* in terms of the radius r makes formula (5.4.6) somewhat rougher, but nevertheless it gives some estimate P_b from above.

For simplicity, we consider a spring less single-mass vibro-impact mechanism. A point mapping of the plane into itself with random changes in R is written in the form:

$$\begin{aligned} O = {} & y_n(\tau_{n+1} - \tau_n) - \cos\tau_{n+1} + \cos\tau_n - (\tau_{n+1} - \tau_n)\sin\tau_n \\ & - (\tau_{n+1} - -\tau_n)^2 p/2 \end{aligned} \tag{5.4.10}$$

$$y_{n+1} = -R_{n+1}\left[y_n + \sin\tau_{n+1} - \sin\tau_n - p(\tau_{n+1} - \tau_n)\right]$$

where y_n the value of velocity immediately after the n-th impact, and p is the ratio of the weight of the striker to the amplitude of the driving force.

The system is characterized by two parameters P and R. The coordinates of the fixed point in the deterministic case are:

$$\sin\tau_0^* = \frac{R-1}{1+R}\pi np$$

$$y_0^* = \frac{2R}{1+R}\pi np \tag{5.4.11}$$

The stability region of single-impact periodic motions is in this case a simply connected domain D_{11}^1. We take some value $p = p_c = 0.6$ and build the area of attraction E_n^j of corresponding fixed points for various $R \in D_{11}^1$.

Figures 5.6, 5.7, 5.8, 5.9, and 5.10 present in the plane of the initial values the attraction regions of fixed points for a fixed value $P = 0.6$ and various values of the coefficient of restitution R equal to 0.31; 0.33; 0.38; 0.40. From these figures, it can be seen that the regions of attraction are non-simply connected. The largest of these areas E_n^j located in the vicinity of a fixed point M_k^*.

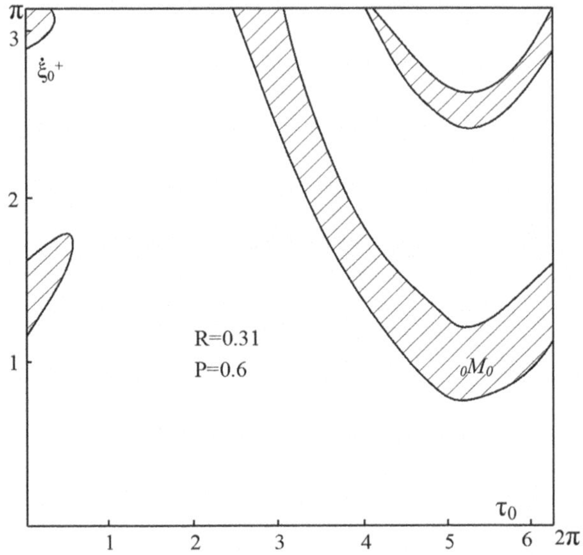

Fig. 5.6 Area of attraction of a fixed point corresponding to a single-impact periodic motion for $R = 0.31, P = 0.6$

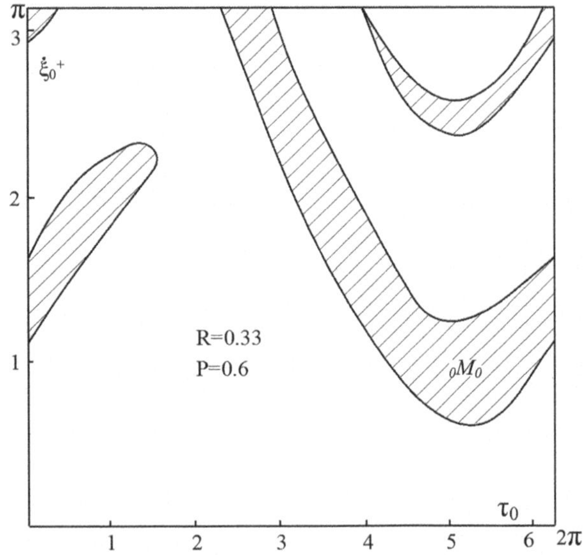

Fig. 5.7 Area of attraction of a fixed point corresponding to a single-impact periodic motion for $R = 0.33, P = 0.6$

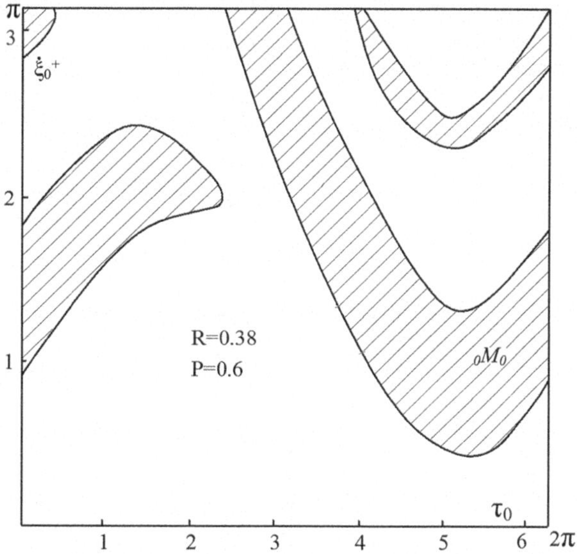

Fig. 5.8 Area of attraction of a fixed point corresponding to a single-impact periodic motion for $R = 0.38$, $P = 0.6$

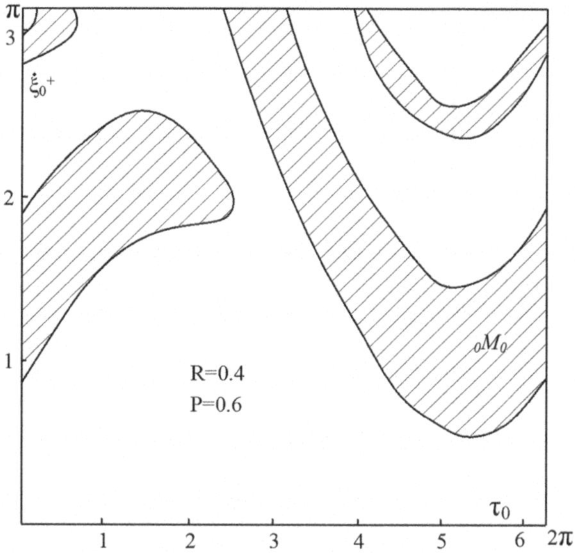

Fig. 5.9 Area of attraction of a fixed point corresponding to a single-impact periodic motion for $R = 0.4$, $P = 0.6$

Linearizing (5.4.7) in the vicinity of periodic motion and using the condition for stability of a fixed point, we obtain the following relations (see formulas (5.1.25) and (5.1.26)) $\Delta\tau_m = T\sum_{v=-\infty}^{n-1}\zeta_{n,v}\Delta R_{v+1}$

$$\Delta y_m = \frac{TP}{1+R}\sum_{v=-\infty}^{n-1}\zeta_{n,v}[m\Delta R_{v+1} - \Delta R_{v+2}] \quad (5.4.12)$$

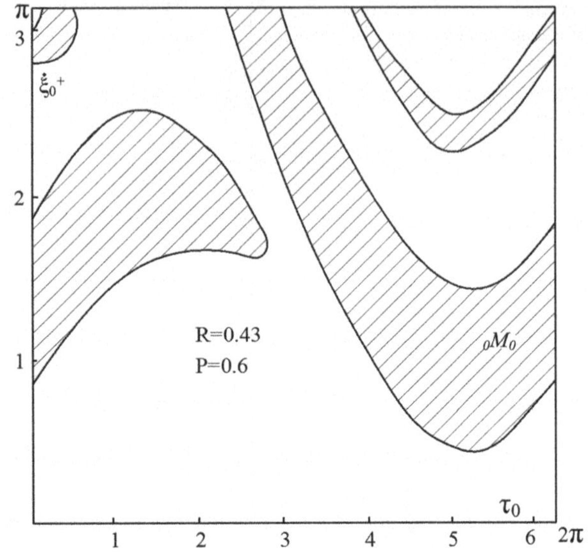

Fig. 5.10 Area of attraction of a fixed point corresponding to a single-impact periodic motion for $R = 0.43$, $P = 0.6$

In our case, random function ξ^2 will be written in the form:

$$\xi^2 = T^2\left\{\left[\sum_{\nu=-\infty}^{n-1} \zeta_{n,\nu}\Delta R_{\nu+1}\right]^2 + \frac{1}{(1+R)^2}\left[\sum_{\nu=-\infty}^{n-1} \zeta_{n,\nu}(m\Delta R_{\nu+1} - \Delta R_{\nu+2})\right]^2\right\} \tag{5.4.13}$$

since $\xi^2 > 0$, then Chebyshev inequality for determining the upper bound P_b can be rewritten as [20]

$$P_b\{\xi^2 > a^2\} \leq \frac{M\xi^2}{a^2} \tag{5.4.14}$$

Assuming further that random changes in R delta are correlated, we obtain

$$\begin{aligned} M\xi^2 &= \frac{T^2\overline{\Delta R^2}}{\left(1-R^2\right)\left(1-Z_1^2\right)\left(1-Z_2^2\right)} \\ &\quad \left\{1+R^2+\frac{P}{(1+R)^2}\left[\left(1+m^2\right)\left(1+R^2\right)-2m(Z_1++Z_2)\right]\right\} \\ &= K(P,R)\overline{\Delta R^2} \end{aligned} \tag{5.4.15}$$

where Z_1,Z_2 are the roots of characteristic Eq. (5.1.24).

Thus, the probability that a random function ξ^2 will leave a circle of radius r^2 and will be less $K(P,R)\overline{\Delta R^2}$ will be written in the form

Table 5.1 Results of calculating the probability of exit P_b depending on the value of the coefficient of recovery of the velocity upon impact R

R	0.31	0.33	0.38	0.4
$K(P, R)$	0.841	0.065	0.0567	0.0784
r	0.1	0.276	0.326	0.320
r^2	0.01	0.076176	0.106276	0.1024
$K(P, R)r^{-2}$	8411.567	648.241	567.49416	783.82109
P_b	0.841	0.065	0.0567	0.0784

$$P_b\{\xi^2 \geq r^2\} \leq \frac{K(P, R)\overline{\Delta R^2}}{r^2}$$

Table 5.1 shows the values of r, $K(P, R)$, P_b for $\overline{\Delta R^2} = 0.01$ for various R and fixed P.

From this table, it can be seen that the probability of exit, as expected, increases with approaching the boundaries of the stability region. Inside the stability region, there is an optimal parameter value for which the probability of exit is minimal (in our case, $R = 0.38$).

5.5 Digital Simulation of Motion of Vibro-hammer with Randomly Changing Coefficient of Restitution

In order to verify the theoretical results and obtain results related to small random changes of R, computer simulation of the motion of a single-mass vibro-hammer with a fixed limiter was carried out with random deviations of the restitution coefficient.

The motion of the vibro-impact mechanism was simulated under the assumption that upon impact the restitution coefficient changes randomly and takes values $R = R_0$ with probability ρ_0, $R = R_0 - \varepsilon -$ with probability ρ_1 and $R = R_0 + \varepsilon$—with probability $\rho_{5.5.2}$ Herewith, $\rho_0 + \rho_1 + \rho_2 = 1$, and ρ_0 larger ρ_1, $\rho_{5.5.2}$.

The modeling process was constructed as follows: for an arbitrary set of system parameters belonging to the stability region, a multiple (N times) construction of the point mapping was carried out according to formula (5.1.22).

If we assume that the distribution of R is equally probable, then the choice of the value R on the n-th impact is reduced to determining the membership of the n-th random number, which is obtained from the random number generator uniformly distributed from 0 to 1 to one of three segments $a_1 = [0, \rho_1]$, $a_2 = [\rho_1, \rho_0]$, $a_3 = [\rho_0, 1]$. If a random number $\alpha_n \in a_k (k = 1, 2, 3)$, then the value of the coefficient of restitution was taken equal $R_0 - \varepsilon$, R_0 and $R_0 + \varepsilon$, respectively. The initial conditions for constructing N times the repeated point mapping were taken equal to the coordinate of the fixed point corresponding to a given set of parameters.

The process of constructing N times the repeated point mapping described above is called the construction of one row of the mN-dimensional matrix A_{mN}. After constructing one row of the matrix, all the necessary statistical characteristics were calculated. In addition, when constructing the matrix row A_{mN}, impacts were recorded at which the system "is lost" due to this mode. We assume that the vibro-impact mechanism "is lost" (exited the mode) if at some moment of impact $k > 1$ $(k \le N)$ the inequality holds:

$$\left|\dot{\xi}_0^* - \dot{\xi}_k\right| > \varepsilon \tag{5.5.1}$$

where $\dot{\xi}_0^*$ isthe value of pre-impact velocity, determined from the condition of periodicity in the deterministic case, $\dot{\xi}_k$ is the value of velocity in "k" impact moment for random changes of R, and here, ε is small, predetermined value.

After complete processing of the first row of matrix A_{mN}, the valuesy$_1 = -R_0$ $\dot{\xi}_0^*$ and $\tau_1 = \tau_0$ were taken for the starting point in constructing the next row, and constructing the next row is similar to the described above process. Thus, m rows of the matrix A_{mN} were constructed. Upon constructing the matrix elements A_{mN}, all statistical characteristics were calculated for the same numbers of impacts, i.e., column characteristics A_{mN}. By the nature of the change in statistical characteristics, condition that for all impact numbers greater than j, the statistical characteristics depend little on the impact number. This means that we enter the stationary mode.

After determining such an impact number j, at which the statistical characteristics are little dependent on the following impacts, one can compare theoretical results with experimental ones.

As a result of the experiments, it turned out that random deviations of the coefficient of restitution R in the range from -0.05 to 0.05 can be considered small. With large deviations $|\Delta R| > 0.05$, theoretical results do not coincide well with the experiment. However, for $0.05 < |\Delta R| < 0$, theoretical results differ from experimental within 20–50% depending on the system parameters.

Digital modeling was carried out for the coefficient of restitution $R = R_0 = 0.6$, which corresponds to the impact of steel against steel.

From experimental results, it follows that at $\varepsilon = 0.001$ and $\varepsilon = 0.005$, all statistical characteristics coincide with the theoretical ones within the accuracy limits of the experiment. So, for these deviation values R, the statistical characteristics can be calculated by the formulas given in Sect. 5.1.

Let us consider the case when $\varepsilon = 0.05$ in detail. Assume that the coefficient of restitution of impact velocity R varies randomly and takes the values 0.55, 0.6, 0.65 with probabilities $\rho_1 = 0, 15$, $\rho_0 = 0, 7$ and $\rho_2 = 0, 15$, respectively. Suppose an equiprobable distribution of R, i.e., the distribution function equals 1. Let the parameter values be as follows $\xi_0 = 1.0$, $\lambda^{-1} = 1.5$. A square matrix $m = 170$, $N = 200$ was constructed in the experiment.

Figure 5.11a–c shows the dependences of dispersion deviation over the impact time, dispersion deviation of the pre-impact velocity versus impact number ($1 \le N \le 200$) when averaging over the columns from the first to 100, 150, 170 rows, respectively. The system output to stationary mode occurs for 20–50 impacts. When calculating $d\tau_n$ and $d\dot{\xi}_n$ according to the formulas given in Sect. 5.1, it turned out that the discrepancy with the experimental data is insignificant.

Figure 5.12 presents the graph of average velocity and average impact time versus impact number when averaging the last row of the matrix.

During the simulation, the coincidence with the theoretical values of the average pre-impact velocity found in Sect. 5.3 was verified. It turned out that $\overline{\dot{\xi}}_{э} = 1.4 \times 10^{-3}$

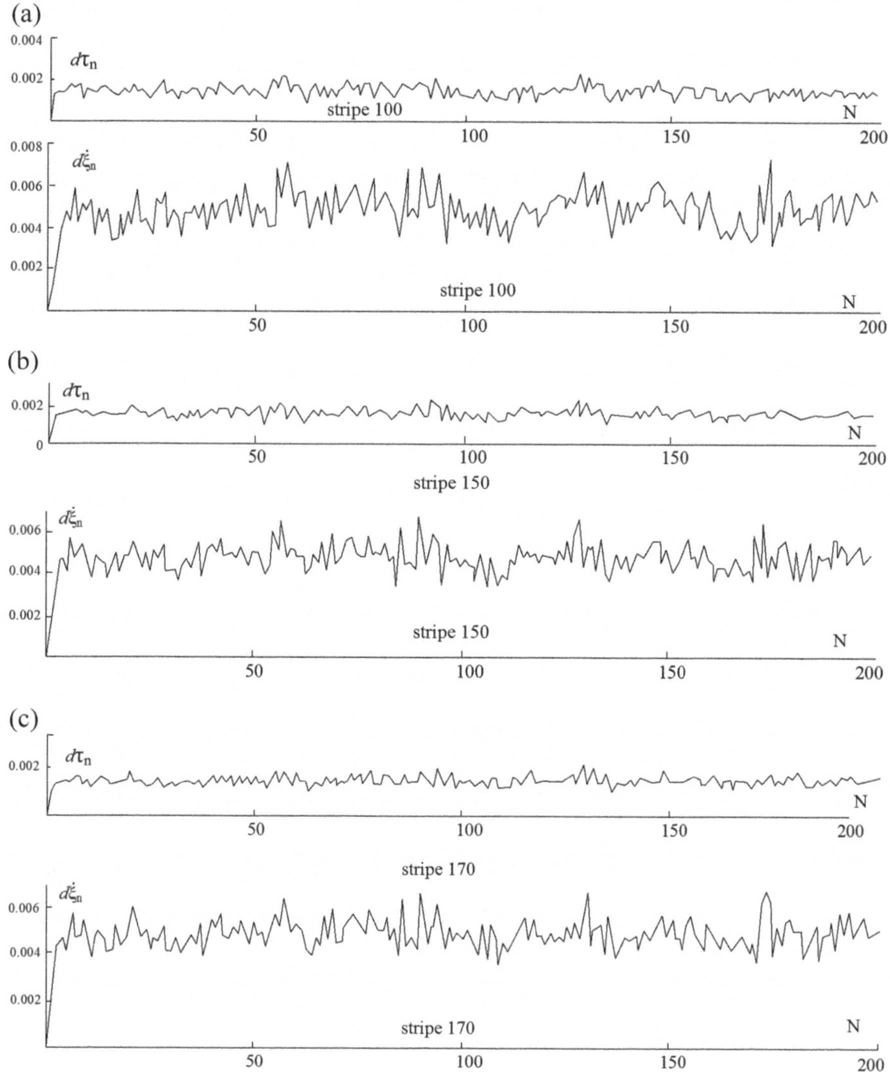

Fig. 5.11 Dependences of the variance of the deviation over impact time and the variance of the deviation of pre-impact velocity depending on the impact number ($1 \leq N \leq 200$) are given when averaging over columns from the first to 100, 150, 170 rows, respectively

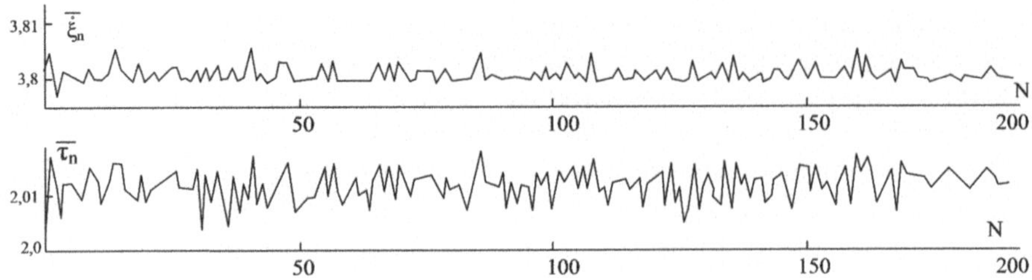

Fig. 5.12 Presents the graph of average velocity and average impact time versus impact number when averaging the last row of the matrix

while $\bar{\xi}_\tau = 1.32 \times 10^{-3}$, where $\bar{\xi}_э$ is the average value of the velocity obtained during the experiment; $\bar{\xi}_\tau$ is the average velocity determined by the formulas Sect. 5.3. The relative error is 5%.

In matrix nodes A_{mN} number was also stored (i, j), in which the system failed. So, if at some moment in the irowat the j-th impact malfunction is observed to occur that $a_{ij} = 1$, otherwise $a_{ij} = 0$. Further, in each column, the summation of all elements took place, i.e., the magnitude was determined

$$\zeta_i = \sum_{j=1}^{m} a_{ij} \tag{5.5.2}$$

which corresponds to the total number of malfunctions at the i-th impact.

Using ζ_i, we can determine the experimental probability of a system failure on an arbitrary i-th impact in the form

$$P_{iэ} = \frac{\zeta_i}{m} \tag{5.5.3}$$

We can calculate ζ_i only if the value $\Delta = \tau_{e+1} - \tau_e$, $1 \le e \le N - 1$ will be less than a certain specified value M. This condition physically corresponds to a transition from a single-impact mode with some i-th impact to single-impact multiple mode.

At $\varepsilon = 0.05$, such malfunctions were not observed.

Figure 5.13 shows a graphical dependency $P_{iэ}$ on impact number N. It is clearly seen that the probability of exit increases from zero to a certain value with impact increasing and then begins to fluctuate in the vicinity of a certain value, which can be taken as the probability of malfunction. In our case, this probability of failure is approximately 0.15.

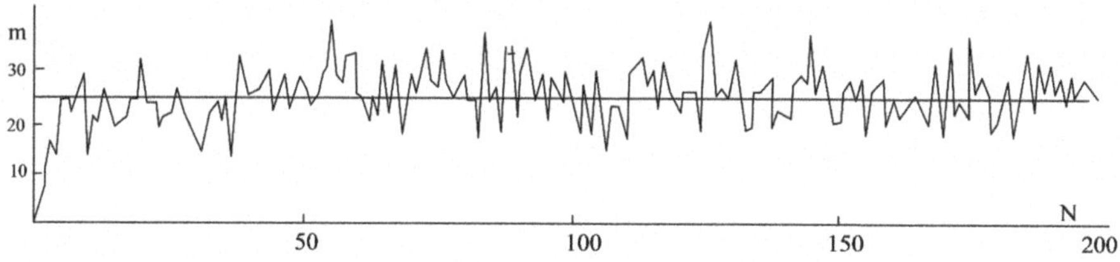

Fig. 5.13 Graphical dependence of P_{ie} on the impact number N is shown

References

1. Bespalova, L.V.: To the theory of vibro-impact mechanism. Izvestia Acad. Sci. USSR, OTN **5** (1957)
2. Goebel, R., Sanfelice, R.G., Teel, A.R.: Hybrid dynamical systems. IEEE Control Syst. Mag. **29**(2), 28–95.3 (2009)
3. Rytov, O.M.: On the theory of fluctuations in self-oscillating systems with piecewise linear characteristics. Izv. High. Academician, Radiophys. **2**(1) (1959)
4. Bykhovsky, I.I., Dorohova, A.D., Zaretsky, L.B., Lukomsky, S.I.: On some periodic motions and the structure of the phase space of a shock-oscillatory system with a constant restoring force. Izv. USSR Acad. Sci. OTV, Mech. Eng. (2) (1964)
5. Sadekov, R.K.: On the issue of fluctuations in piecewise linear self-oscillatory systems. Izv.vyssh.uch.zav., Radiophys. **5.3**(5) (1960)
6. Khasminsky, R.Z.: Stability of systems of differential equations under random perturbations of their parameters. Ed. Sci., M. (1969)
7. Nevelson, M.B., Khasminsky, R.Z.: Problems of information transmission. **2**(3) (1966)
8. Nevelson, M.B.: On the operation of a linear system with small random excitations of its parameters. PMM **31**(3) (1967)
9. Nevelson, M.B., Khasminsky, R.: On the stability of a linear system under random excitations of its parameters. PMM (2) (1966)
10. Nevelson, M.B.: Some remarks on the stability of a linear stochastic system. PMM (6) (1966)
11. Khasminsky, R.Z.: On stability in the first approximation of stochastic systems. PMM **3**(6) (1967)
12. Katz, I.Y.: On stability in first approximation of a system with random delay. PMM **31**(3) (1967)
13. Brusin, V.A., Tai, M.J.I.: Absolute stochastic stability. Izv. High. uch. Head, Radiophys. **10**(7) (1967)
14. Kolomeets, V.G., Korenvsky, D.G.: On the stability of linear systems with random excitations. Appl. Mech. **3**(8)
15. Kats, I.Y., Krasovsky, N.N.: On the stability of systems with random parameters. PMM **27**(5) (1960)
16. Weidenhammer, F.: Stabilitat sbedingungen fur Schwingungssysteme mit zufalliger Parametererregung durch weibes Rauschen, Ingen.-Archiv. **35**(1), 1–9 (1966)
17. Neimark, Y.I., DAN USSR **127** (1959)
18. Goebel, R., Sanfelice, R.G., Teel, A.R.: Hybrid Dynamical Systems Modeling, Stability, and Robustness. Princeton University Press (2012)
19. Hayashi, T.: Forced Oscillations in Nonlinear Systems, Ed. I.L., M. (1957)
20. Goebel, R., Teel, A.R.: Solutions to hybrid inclusions via set and graphical convergence with stability theory applications. Automatica **42**(4), 573–587 (2006)

The manufacturer's authorised representative in the EU is Springer Nature Customer Service Centre GmbH, Europaplatz 3, 69115 Heidelberg, Germany. If you have any concerns regarding our products, please contact ProductSafety@springernature.com

Printed and bound by CPI Group (UK) Ltd, Croydon, CR0 4YY
01/10/2026
02225982-0014